Time-Dependency in Rock Mechanics and Rock Engineering

ISRM Book Series
Series editor: Xia-Ting Feng
Institute of Rock and Soil Mechanics, Chinese Academy of Sciences, Wuhan, China

ISSN : 2326-6872
eISSN: 2326-778X

Volume 2

International Society for Rock Mechanics

ISRM

Time-Dependency in Rock Mechanics and Rock Engineering

Ömer Aydan

Department of Civil Engineering and Architecture, University of the Ryukyus, Nishihara, Okinawa, Japan

CRC Press
Taylor & Francis Group
Boca Raton London New York

CRC Press is an imprint of the
Taylor & Francis Group, an **informa** business
A BALKEMA BOOK

Published by:
CRC Press/Balkema
P.O. Box 447, 2300 AK Leiden, The Netherlands
e-mail: Pub.NL@taylorandfrancis.com
www.crcpress.com – www.taylorandfrancis.com

First issued in paperback 2021

ISBN-13: 978-1-03-209728-2 (pbk)
ISBN-13: 978-1-138-02863-0 (hbk)

Publisher's Note
The publisher has gone to great lengths to ensure the quality of this reprint but points out that some imperfections in the original copies may be apparent.

Visit the Taylor & Francis Web site at
http://www.taylorandfrancis.com

and the CRC Press Web site at
http://www.crcpress.com

Typeset by MPS Limited, Chennai, India

Library of Congress Cataloging-in-Publication Data

Table of contents

About the author

Born in 1955, Professor Aydan studied Mining Engineering at the Technical University of Istanbul, Turkey (B.Sc., 1979), Rock Mechanics and Excavation Engineering at the University of Newcastle upon Tyne, UK (M.Sc., 1982), and finally received his Ph.D. in Geotechnical Engineering from Nagoya University, Japan in 1989. Prof. Aydan worked at Nagoya University as a research associate (1987–1991), and then at the Department of Marine Civil Engineering at Tokai University, first as Assistant Professor (1991–1993), then as Associate Professor (1993–2001), and finally as Professor (2001–2010). He then became Professor of the Institute of Oceanic Research and Development at Tokai University, and is currently Professor at the University of Ryukyus, Department of Civil Engineering & Architecture, Nishihara, Okinawa, Japan. He has furthermore played an active role on numerous ISRM, JSCE, JGS, SRI and Rock Mech. National Group of Japan committees, and has organized several national and international symposia and conferences. Professor Aydan has received the 1998 Matsumae Scientific Contribution Award, the 2007 Erguvanlı Engineering Geology Best Paper Award, the 2011 Excellent Contributions Award from the International Association for Computer Methods in Geomechanics and Advances, the 2011 Best Paper Award from the Indian Society for Rock Mechanics and Tunnelling Technology and was awarded the 2013 Best Paper Award at the 13th Japan Symposium on Rock Mechanics and 6th Japan-Korea Joint Symposium on Rock Engineering. He was also made Honorary Professor in Earth Science by Pamukkale University in 2008 and received the 2005 Technology Award, the 2012 Frontier Award and the 2015 Best Paper Award from the Japan National Group for Rock Mechanics.

Acknowledgements

The author sincerely acknowledges Prof. Xia-Ting Feng for inviting the author to contribute to the ISRM Book Series on "Time Dependency in Rock Mechanics and Rock Engineering". The content of this book is an outcome of the studies carried out by the author at Newcastle upon Tyne University (UK), Nagoya University, Tokai University and University of the Ryukyus in Japan for more than 3 decades. The author would like to thank Prof. Ian Farmer, formerly with Newcastle upon Tyne University (UK), Emeritus Prof. Dr. Toshikazu Kawamoto of Nagoya University, Prof. Dr. Tomoyuki Akagi and Prof. Dr. Takahi Ito of Toyota National College of Technology, Prof. Dr. Takashi Kyoya of Tohoku University, Prof. Dr. Takafumi Seiki of Utsunomiya University, Japan, Prof. Dr. Reşat Ulusay of Hacettepe University and Prof. Dr. Halil Kumsar of Pamukkale University, Turkey for their guidance, help and suggestions at various stages of his studies quoted in this book. The author also acknowledges his former students, particularly J.G. Chen of Nagoya University and Mitsuo Daido and Yoshimi Ohta of Tokai University for their help during experiments and computations reported in this study. The author would also like to thank Alistair Bright, Acquisitions Editor at CRC Press/Balkema for his patience and collaborations during the preparation of this book and Ms. José van der Veer, production editor, for the great efforts to produce this book.

Chapter 1

Introduction

Long term response and stability of rock engineering structures such as tunnels, underground openings, slopes and stone-made cultural assets have been receiving great attention since early times. Furthermore, the stability of the room and pillar mines during exploitation and after abandoning is also of great concern (Figure 1.1). There are many causes affecting their long term response and stability, such as sustained loading with or without additional loads resulting from various sources of blasting, machine vibration and earthquakes, freezing and thawing, and weathering due to physical and chemical actions of the percolating fluids.

Figure 1.1 Collapse of some underground openings involving time-dependent characteristics of rocks.

Figure 1.2 Some nuclear power plants and underground nuclear waste disposal caverns.

The nuclear waste disposal projects in countries utilizing nuclear energy and/or nuclear weapons require the consideration of a very long time span of at least 10,000 years for the assessment of response and stability of underground disposal facilities in rock. Particularly, the nuclear waste disposal projects involve very complex interactions of thermal, hydrological, diffusive and mechanical phenomena (Figure 1.2).

In Geo-science, it is well known that faults and the earth's crust show creep-like responses. Some case histories from San Andreas and North Anatolian fault are well documented. The crustal deformation response before the 2003 Miyagi-Hokubu earthquake in Japan was very similar to what is observed in creep experiments (Figure 1.3). There are also many experimental studies on the creep and relaxation behaviour of crustal rocks under different temperature regimes. The creep test is an experiment carried out under sustained loading condition. The load is generally applied in a step-like fashion. For time dependent or rate-dependent characteristics of rocks, there are different methods such as rate-dependent experiments besides creep tests. In actual sense, there are some correlations between these two different types of experiments.

This book is concerned with time-dependency in rock mechanics and rock engineering, whose spectrum is very wide as mentioned above. While the term "time-dependency" involves time-dependent behaviour/rate-dependent behaviour of rocks in a conventional sense, this book attempts to cover the spectrum as much as possible

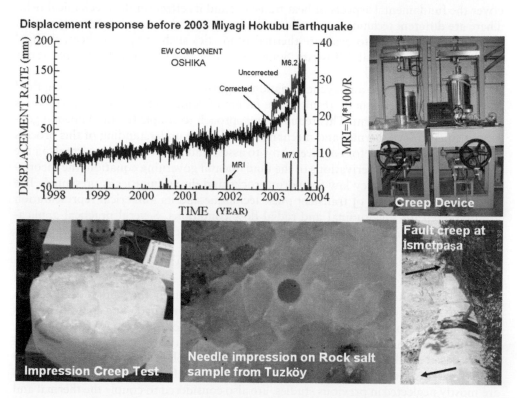

Figure 1.3 Examples of creep of rocks and faults and testing devices.

including coupled processes of thermal, hydrological and diffusions in rocks. This book specifically deals with the following topics.

Chapter 2: As all rocks exhibit time-dependent behavior, the long term response and stability of rock engineering structures are of great importance and the engineers must know how to deal with this issue. This chapter covers many aspects of time/rate dependency of rocks and associated engineering issues.

Chapter 3: Degradation of mechanical properties of soft rocks due to absorption and desorption of water as a result of cyclic drying-saturation process is another important issue in rock mechanics and rock engineering in the long term. This process involves moisture variation, which results in volumetric changes, causing their cracking and decomposition. The scientists and engineers involved with rocks must know this process and how to assess the performance of structures in such rocks in the long term. This chapter describes the fundamentals of this phenomenon, its mechanical modeling and some specific applications to actual problems.

Chapter 4: Heat transport in rocks is another important issue for dealing with the long term performance of structures of great importance as well as geothermal field exploitation. Furthermore, the understanding of mechanical property variations in relation to temperature fluctuations and their effect on rock engineering structures are also of great significance for rock engineers and scientists. This chapter is intended to cover the fundamental aspects of heat transport and its effect on the mechanical field. There are different techniques to evaluate thermal properties of rocks. A very practical procedure is proposed to evaluate thermal properties such as specific heat, thermal conductivity and thermal diffusivity. Some specific examples of practical applications are also presented.

Chapter 5: Seepage and cyclic variations of groundwater as free percolating fluid in rock mass may eventually result in the failure of some rock engineering structures in the short and long term. This requires an approach to couple the fundamental governing equations of thermal and mechanical fields. The understanding of this process should be of great value for engineers how to assess and monitor their structures in the long term. Besides the derivations of the fundamental governing equation, the theoretical background of Darcy law for porous rocks and discontinuities is presented. The theoretical background of transient pulse technique and its numerical representation are presented for longitudinal and radial flow conditions. Several practical applications of seepage in rock mass in relation to some specific rock engineering structures are presented.

Chapter 6: The nuclear-waste disposal issue in rock engineering is a very challenging problem for scientists and engineers of rocks and it involves very sophisticated interaction of fundamental governing equations. Particularly the studies dealing with radioactive waste disposal have been concerned with thermo-hydro-mechanical aspects of the phenomenon as well as the diffusion phenomenon of the radioactive substances. Since the diffusion phenomenon is quite important in the long term, a mechanical model based on the mixture theory is described to couple thermal, hydrological and diffusion fields in this chapter. In the presented model, Duffour and Soret effects, which are mostly neglected in previous studies, are also considered to couple the thermal and diffusion fields. Then a finite element formulation of the derived theoretical model is given and a series of numerical analyses carried out for the simulation of laboratory tests is presented. Furthermore, some parametric studies are performed to investigate the coupling effects of Duffour and Soret effects on thermal and diffusion fields.

Chapter 7: The near-field disposal of nuclear wastes in rock is associated with thermo-hydro-mechanical aspects of the phenomenon. While the heat is emitted from the nuclear waste contained in canisters, the in-situ stress in rock mass and seepage of ground water is of great significance for the stability of the waste disposal sites. This chapter describes a thermo-hydro-mechanical model based on the mixture theory to couple thermal, hydrological and mechanical fields. Then a finite element formulation of the derived theoretical model is presented and a series of numerical analyses was carried out for simulating some laboratory experiments of this phenomenon.

This book presents theoretical formulations, some experimental techniques, numerical formulations and examples of applications. If this book is used as a textbook for educational purposes, Chapters 2 to 5 may be used for both undergraduate and graduate courses. Chapters 6 and 7 would be for graduate courses, particularly.

Time-dependent (rate-dependent) behaviour of rocks

2.1 INTRODUCTION

Long term response and stability of rock engineering structures such as tunnels, underground openings, slopes and stone-made cultural assets have been receiving great attention since early times. Furthermore, the stability of the room and pillar mines during exploitation and after abandoning is also of great concern (Mottahed & Szeky, 1982; Doktan, 1983). There are many causes affecting their long term response and stability such as sustained loading with or without additional loads resulting from various sources of blasting, machine vibration and earthquakes, freezing and thawing, weathering due to physical and chemical actions of the percolating fluids.

The nuclear waste disposal projects in countries utilizing nuclear energy and/or nuclear weapons require the consideration of very long time span of at least 10,000 years for the assessment of response and stability of underground disposal facilities in rock. Particularly, the nuclear waste disposal projects involve very complex interactions of thermal, hydrological, diffusive and mechanical phenomena.

Most of cultural assets from previous civilizations are structures made of various stones. The deterioration of these cultural assets due to natural causes is a serious problem to be dealt with. In addition to atmospheric agents and percolating fluids, long term sustained loading causes deformation and instability of those cultural assets.

In Geo-science, it is well known that faults and the earth's crust show creep-like responses. Some case histories from San Andreas and North Anatolian fault are well documented. The crustal deformation response before the 2003 Miyagi-Hokubu earthquake in Japan was very similar to what is observed in creep experiments. There are also many experimental studies on the creep and relaxation behaviour of crustal rocks under different temperature regimes.

Time-dependent behaviour of rocks has been experimentally studied since early times (see the textbook by Jaeger & Cook (1979) and Cristescu & Hunsche (1998) for details). The extensive experimental studies were performed on halides (rocksalt or halite and potash) as they have been considered good sealing rocks for the containment and disposal of nuclear wastes. The experiments were mainly carried on rock salts subjected to creep loading conditions under different constant temperature regimes (i.e. Wawersik, 1983; Hunsche, 1992). Almost all experiments were carried out under compressive uniaxial and/or triaxial loading conditions. In some of experiments, the healing process of rock salts was also studied.

Creep characteristics of rocks are very important for assessing the long term stability of rock engineering structures. A series of experiments for the creep characteristics of soft rocks was undertaken by the author at Tokai University and Toyota National College of Technology using uniaxial compression, Brazilian creep and impression creep testing methods. The author reports the results of experiments carried out on Oya tuff. Then a series of numerical studies were carried out to investigate the stress-strain fields induced in each type of experimental technique and their possible correlations. In the final part of the report, several examples of the utilization of creep characteristics of soft rocks for assessing the long term performance and stability of rock engineering structures are presented and discussed.

For time dependent or rate-dependent characteristics of rocks, there are different methods such as rate-dependent experiments besides creep tests. In actual sense, there are some correlations between these two different types of experiments as discussed in the work of Aydan & Nawrocki (1998).

2.2 CREEP BEHAVIOUR AND TESTING TECHNIQUES

The creep experiments are often used to determine the time-dependent strength and/or time-dependent deformation modulus of rocks. It has often been stated that the creep of rocks does not occur unless the load/stress level exceeds a certain threshold value, which is sometimes defined as the long term strength of rocks (Ladanyi, 1974; Bieniawski, 1970). However, experiments carried on igneous rock (i.e. granite, gabbro etc.) beams by Ito (1991) for three decades show that a creep response definitely occurs even under very low stress levels. The threshold value suggested by Ladanyi (1993) may be associated with the initiation of dilatancy of volumetric strain as illustrated in Figure 2.1. The initiation of dilatancy generally corresponds to 40–60% of the stress

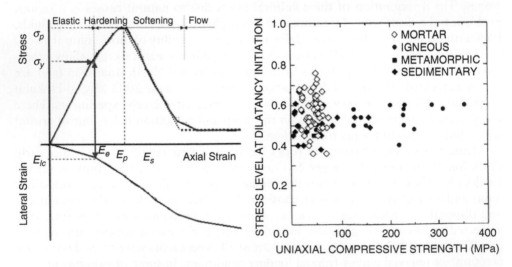

Figure 2.1 Illustration of threshold value for dilation and experimental results for different rocks (arranged from Aydan *et al.,* 1993, 1994).

level and the fracture propagation tends to become unstable when the applied stress level exceeds 70–80% level of the ultimate deviatoric strength for given stress state (Aydan *et al.*, 1994; Hallbauer *et al.*, 1973). Therefore, the behaviour below the threshold generally corresponds to visco-elastic behaviour. Creep threshold according to Ladanyi (1974) corresponds to an elasto-visco-plastic response and it should not be possible to obtain visco-elastic properties directly from measured responses.

The creep responses terminating in failure are generally divided into three stages as shown in Figure 2.2. These stages are defined as primary, secondary and tertiary creep stages. The secondary stage appears to be a linear response in time (but in a real sense, it is not a linear response). On the other hand, the tertiary stage is the stage in which the strain response increases exponentially resulting in the failure of creep behaviour. The modeling of this stage in the constitutive laws is an extremely difficult aspect as it also depends upon the boundary conditions.

2.2.1 Laboratory creep testing devices

Apparatuses for creep tests can be of the cantilever type or the load/displacement-controlled type. Although the details of each testing machine may differ, the features of apparatuses for creep tests are described herein.

(a) Cantilever type testing device

The cantilever-type apparatus has been used in creep tests since early times (i.e. Serata *et al.*, 1968; Akagi, 1976; Farmer, 1983; Ito & Akagi, 2001) (Figure 2.3). It is in practice the most stable apparatus for creep tests because the load level can easily be kept constant with time. The greatest restrictions of this type apparatus are the level of applicable load, which depends upon the length of the cantilever arm and its oscillations during the application of the load. The cantilever-type apparatus utilizing a multi-arm lever overcomes the load limit restrictions (Okada, 2005, 2006). The oscillation is another technical problem to be dealt by the producers of the creep devices. If the load increase is manually done through putting deadweights in some creep testing devices, an utmost care must be undertaken during loading procedure in order to prevent undesirable oscillations.

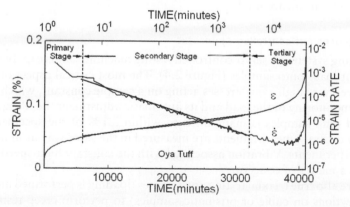

Figure 2.2 Strain and strain rate response of a creep experiment on Oya tuff (Japan).

(b)   (a)

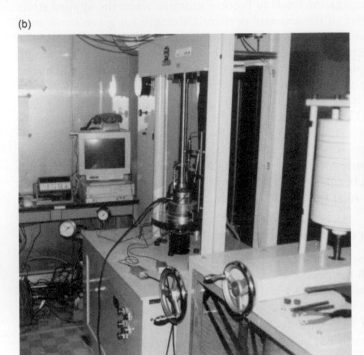

Figure 2.3 Examples of cantilever type creep apparatuses: (a) Single arm cantilever and (b) multi-lever.

Deformation and strain measurements can be taken in several ways. The simple approach is to utilize a couple of LVDTs. When a triaxial creep experiment is carried out, the LVDTs may be fixed onto the sample and inserted into the triaxial chamber. In such a case, special precautions are necessary for the accurate measurement of displacements. Strain gauges may be used; however, the strain gauges glued onto samples are required to be capable of measuring strain over a long period of time without any debonding. For lateral deformation or strain measurements, diametric or circumferential sensors are used.

(b) Load/displacement controlled apparatus

Loading testing system is a servo-controlled testing machine that is capable of applying high constant loads onto samples (Figure 2.4). The most critical aspect of this experiment is to keep very high axial stresses acting on a sample constant, which will require continuous monitoring of the load and its automatic adjustment (i.e. Peng, 1973). The load applied onto samples is maintained to within ±1% of the specified load. The deformation or strain measurements are measured in the same way as in the cantilever type creep experiments. Vibration associated with the constant high-speed closed-loop operation is a matter of concern.

There are also true triaxial testing apparatuses (loading is performed independently in three directions on cubic or prismatic samples) to perform creep tests under true triaxial stress conditions (Serata *et al.*, 1968; Adachi *et al.*, 1969). Three principal

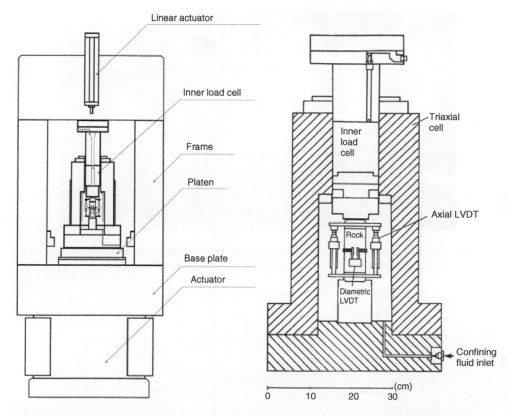

Figure 2.4 Load/displacement controlled apparatus (from Ishizuka *et al.*, 1993).

stresses can be controlled independently in such triaxial testing apparatuses. New technologies make such tests to be performed much easier.

Creep tests under direct shear stress condition on rocks, discontinuities and interfaces are also carried out using a servo-control loading system (i.e. Amadei & Curran, 1982; Aydan *et al.*, 1994, 2016; Voegler *et al.*, 1998; Larson & Wade, 2000). Figure 2.5 shows the multi-purpose dynamic shear-testing machine with an ability to perform creep tests on rocks, discontinuities and interfaces at the University of the Ryukyus. The device was originally developed for conventional direct shear creep test and cyclic shear tests and has been recently upgraded to perform dynamic shear testing (Aydan *et al.*, 1994, 2016).

2.2.2 Laboratory creep tests

(a) Uniaxial compression creep tests

Creep tests on Oya tuff carried out by Ito & Akagi (2001) under dry conditions are plotted in Figure 2.7. As noted from Figure 2.6, some of responses terminate with failure while the others become asymptotic to certain strain levels, depending on the

Figure 2.5 Multi-purpose dynamic shear-testing machine with a capability to perform creep tests on rocks, discontinuities and interfaces at the University of the Ryukyus.

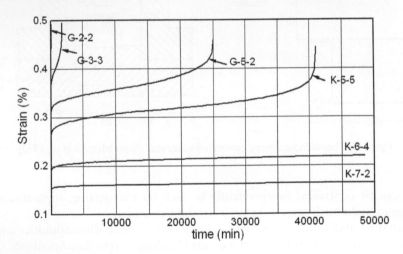

Figure 2.6 Uniaxial compression creep response of Oya tuff under dry condition (modified from Ito & Akagi 2001).

applied stress ratio (SR), which is defined as the ratio of applied stress to the short term strength. The responses terminating in failure are generally divided into three stages as shown in Figure 2.2. The transitions from the primary stage to the secondary stage and from the secondary stage to the tertiary stage are generally determined from the deviation of a linearly decreasing or increasing strain rate plotted in a logarithmic time space. Generally, it should, however, be noted that strain data must be smoothed before its interpretation. Direct derivation of strain data containing actual responses as well as electronic noise may produce entirely different results.

When rocks have water absorption ability, their strength tends to decrease compared with that under dry condition (i.e. Aydan, 1993; Aydan & Ulusay, 2002, 2013). Particularly, the strength of soft rocks like tuffs decreases drastically and the strength

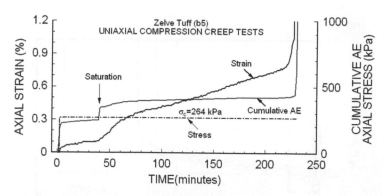

Figure 2.7 Responses of initially dry later saturated tuff samples from Zelve (b5) during uniaxial compression creep tests (arranged from Ito *et al.*, 2008).

reduction is generally greater than 60%. In some cases, soft rocks may disintegrate upon water absorption and the resulting strength reduction may be up to 100%. Several researchers investigated the effect of saturation on uniaxial compression creep tests (i.e. Ito *et al.*, 2008; Okubo & Chu, 1994; Okubo *et al.*, 2005; Aydan *et al.*, 2013). Figure 2.7 shows an example of creep response of a tuff sample from Zelve, Cappadocia (Turkey). The sample was initially subjected to a creep loading at a level of about 16% of its uniaxial compressive strength under dry conditions. The sample was fully saturated 40 minutes after the start of the creep test. The stress ratio becomes about 95% of the uniaxial compressive strength under saturated condition. As the stress ratio increased, the sample failed about 190 minutes after the saturation. Creep experiments carried out on tuff samples from Derinkuyu, Avanos and Ürgüp yielded similar results (i.e. Aydan & Ulusay, 2013; Ulusay *et al.*, 2013).

Effect of temperature on creep response of various rocks is investigated by various researchers (i.e. Shibata *et al.*, 2007; Okada, 2005, 2006; Cristescu & Hunsche, 1998; Hunsche & Hampell, 1999). It is well known that the strength of rocks decreases with temperature (i.e. Handin, 1966; Shimada, 1993; Hirth & Tullis, 1994; Brace & Kohlstedt, 1980). Figure 2.8 shows plots of responses during uniaxial compression creep tests on Oya tuff and its failure time determined at different temperatures. As noted from the figure, the creep response is accelerated and the long term uniaxial compression strength of Oya tuff decreases.

(b) Triaxial compression creep tests

Triaxial compression creep experiments are quite limited as compared with uniaxial compression creep experiments due to sophistication of equipments and costs. Nevertheless, there were several attempts to conduct such tests (i.e. Serata *et al.*, 1968; Lockner & Byerlee, 1977; Waversik, 1983; Masuda *et al.*, 1987; Okada, 2005; Ito *et al.*, 1999). Provided that friction angle is not rate-dependent, the stress ratios under triaxial compression creep test are defined in an analogy to that in uniaxial compression creep tests as:

$$SR = \frac{\sigma_1 - \sigma_3}{2c\cos\phi + (\sigma_1 + \sigma_3)\sin\phi} \tag{2.1}$$

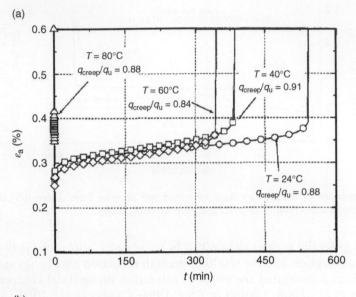

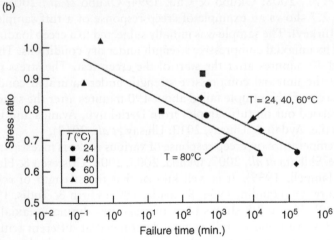

Figure 2.8 (a) Creep response of Oya tuff and (b) Relationship between stress ratio and failure time at various temperatures (arranged from Shibata et al., 2007).

where c, ϕ, σ_1 and σ_3 are cohesion, friction angle and maximum applied and confining stresses, respectively. If friction angle is rate-dependent, the ratio of the applied deviatoric stress to the deviatoric strength is used as stress ratio. However, the experimental results confirm that the rate-dependency of friction angle is negligible according to Aydan & Nawrocki (1998).

Figures 2.9 and 2.10 shows the creep response under a confining stress of 2 MPa and the failure time of compression creep tests under both uniaxial and triaxial compression environment. It is interesting to note that the overall tendency obtained in triaxial creep tests is basically similar to those of uniaxial compression creep tests irrespective of confining pressure.

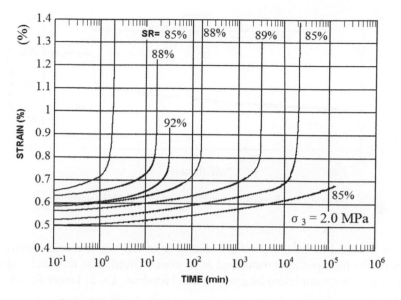

Figure 2.9 Creep response at confining pressure of 2 MPa.

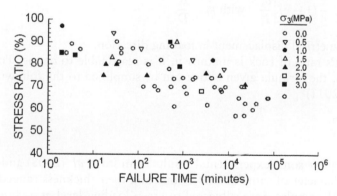

Figure 2.10 Creep failure time of Oya tuff uniaxial and triaxial compression creep tests (arranged from Ito et al., 1999; Shibata et al., 2007; Akai et al., 1979).

(c) Brazilian tensile creep tests

There are not many studies on tensile creep behaviour of rocks using Brazilian creep tests. However, rock may be subjected to tensile stresses in nature such as cliffs with toe erosion and roof layers above underground openings excavated in sedimentary rocks. Aydan *et al.* (2011, 2013), Agan *et al.* (2013) and Ulusay *et al.* (2013) have recently reported some Brazilian creep tests on tuff samples. The tensile strength of the specimen is calculated using the well-known following formula:

$$\sigma_t = \frac{2}{\pi} \frac{P}{Dt} \tag{2.2}$$

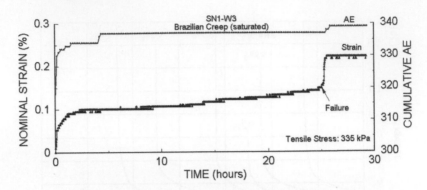

Figure 2.11 Brazilian creep response of SN1-W3 sample.

where P is the load at failure, D is the diameter of the test specimen (mm), t is the thickness of the test specimen measured at its center (mm). The nominal strain of the Brazilian tensile test sample can be given as (see Hondros, 1959; Jaeger & Cook, 1979; for details)

$$\varepsilon_t = 2\left[1 - \frac{\pi}{4}(1 - v)\right]\frac{\sigma_t}{E} \quad \text{with } \varepsilon_t = \frac{\delta}{D} \qquad (2.3)$$

where δ is diametrical displacement in loading direction.

If Poisson's ratio of rock is unknown, it is reasonable to choose Poisson's ratio as 0.25. Thus, the formula given above can be simplified to the following form (i.e. Aydan *et al.*, 2011)

$$\varepsilon_t = 0.82\frac{\sigma_t}{E} \qquad (2.4)$$

Here we quote some experimental results from Ito *et al.* (2008) and Aydan *et al.* (2011). The diameter of samples was 46 mm and their thickness ranged between 14 and 25 mm. All samples were subjected to creep loading level at a chosen period of time under dry conditions. After reaching the ultimate loading level, the samples were saturated. Figure 2.11 shows some of the measured response of a sample in Brazilian creep experiments on Oya tuff. Oya tuff sample numbered SN1-W3 was tested under fully saturated conditions at a stress ratio of 87%. As noted from the figures, acoustic emission occurs at each load increase, simultaneously.

(d) Direct shear creep tests

Direct shear creep tests on rocks, discontinuities and interfaces are also quite rare. Amadei & Curran (1982) performed direct shear creep tests on rock discontinuities. The direct shear tests by Aydan *et al.* (1994, 2016), Voegler *et al.* (1998), Larson & Wade (2000) may be counted in addition the initial tests performed by Amadei & Curran (1982). We present the experimental results by Aydan *et al.* (1994) performed on the interfaces and grouting material in rock anchor systems. Figure 2.12 shows

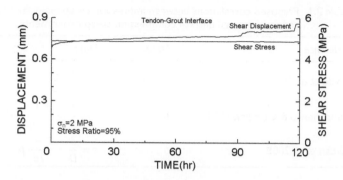

Figure 2.12 Direct shear creep test on tendon-grout interface.

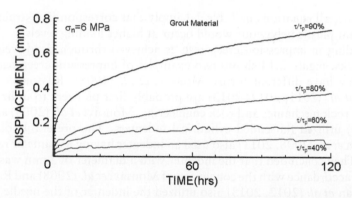

Figure 2.13 Responses of grouting material measured during direct shear creep test at various stress ratios.

the direct shear creep experiment on tendon-grout interface under a normal stress of 2 MPa. The stress ratio was about 95%. The overall response is similar to those of uniaxial and triaxial compression and Brazilian creep tests.

Figure 2.13 shows the creep responses of grouting material of rock anchor systems tested under direct shear condition. The initial instantaneous displacements are subtracted from displacement response for each stress ratio. Similarly, the creep displacement increases as the stress ratio becomes higher.

(e) Impression creep experiments

Impression creep experiments are relatively easy to perform and the capacity of loading equipments is relatively small compared to conventional creep experiments. The critical issue with this technique is the definition of strain and stress, which can be associated with conventional creep experiments. There are several proposals on how to correlate impression creep experiments to conventional creep experiments, which are summarized in Table 2.1 (i.e. Hyde et al., 1996; Timoshenko & Goodier, 1970; Sastry, 2005; Rassouli et al., 2010; Aydan et al., 2011). If applied load is assumed

Table 2.1 Proposed correlations between impression creep experiments and conventional uniaxial compression creep experiments.

Reference	Stress	Strain
Hyde et al., 1980	$\sigma = \eta p$	$\varepsilon = \dfrac{1}{\beta} \cdot \dfrac{\delta}{D}$
Timoshenko & Goodier, 1970	p	$\varepsilon = \dfrac{\delta}{D} = \dfrac{\pi(1-\nu^2)}{4E} p$
Aydan et al., 2008	p	$\varepsilon = \dfrac{\delta}{D} = \dfrac{1+\nu}{2E} p$

to be the same, all equations in Table 2.1 imply that corresponding strains would be smaller so that plastic behaviour would occur at higher loading levels.

The loading in impression creep tests is achieved through dead weights and/or hydraulic jacks. Figure 2.14 shows two examples of impression creep testing device. Indenters may have different forms. Mousavi et al. (2008), Rassouli et al. (2010) and Aydan et al. (2011, 2012, 2013) are probably first pioneers to utilize this index technique in rock mechanics and rock engineering. Mousavi et al. (2008) and Rassouli et al. (2010) utilized flat-ended cylindrical indenters. The preferable diameter was 3 mm. Aydan et al. (2008, 2011) also used an indenter having a diameter ranging from 1 to 3 mm. They concluded that the indenter with a diameter of 3 mm was preferable, which are in accordance with the conclusion of Mousavi et al. (2008) and Rassouli et al. (2010). Aydan et al. (2012, 2013) also utilized the indenter of the needle penetration index test device (Aydan, 2012, 2013; Ulusay et al., 2013).

The experimental results are presented in this subsection using the device shown in Figure 2.14(a) with a flat-ended indenter having a diameter of 3 mm. The device is capable of inducing loads, which is 10 times the applied load at the end of the arm. The device was equipped with a displacement transducer and an acoustic emission sensor. However, electric potential measurement system could be included in the monitoring system under dry condition. Figure 2.15 shows the results of an impression creep test on Oya tuff sample denoted WEZ-4 under saturated condition. The saturated strength of Oya tuff is about 40–50% of that under dry condition and yielding level is expected to be more than 14 MPa. The response becomes stable following the applied nominal pressure of 12.2 MPa. However, the sample fails when the applied pressure is 21 MPa. The stress ratio is about 61% in view of the short term indentation tests.

An impression creep experiment carried out on rocksalt sample from Tuzköy in Cappadocia region of Turkey. The short term and long term properties of this rocksalt was investigated by Özkan et al. (2009) and Özşen et al. (2014) under uniaxial compression creep test. The short term average uniaxial compressive strength of Tuzköy rock salt is about 26.5 MPa. Figure 2.16 shows the response obtained from the impression creep test. The load level was gradually increased in steps up to 85 MPa. In the last three steps, the amplitude of load was decreased to 28 MPa first and increased to designated level greater than the previous state. It was noted that the elastic recovery was very small and the behaviour of rocksalt was almost visco-plastic. Upon unloading at

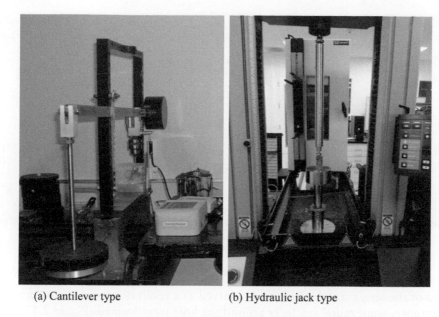

(a) Cantilever type   (b) Hydraulic jack type

Figure 2.14 Two examples of impression creep devices.

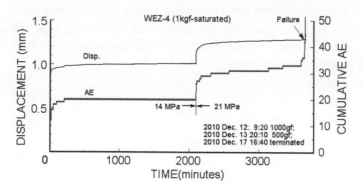

Figure 2.15 Impression creep response of saturated Oya tuff sample denoted WEZ-4.

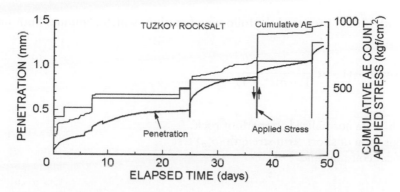

Figure 2.16 Response of Tuzköy rocksalt during the impression creep test.

Figure 2.17 Views of Tuzköy rocksalt sample during and after impression test.

the end of the test, a circular hole was observed as a result of permanent deformation. Furthermore, some radial fractures around the hole were formed (Figure 2.17).

(f) Long term strength of rocks and correlation among various creep tests

The strength of rocks is generally assumed to be hardening type. However, it is well known that the long term strength ($\sigma_a(t)$) of rocks decreases with time and it is expressed in the following forms

Aydan *et al.* (1996)

$$\frac{\sigma_a(t)}{\sigma_{co}} = \alpha + (1 - \alpha)e^{-b(t^* - 1)} \tag{2.5}$$

Aydan & Nawrocki (1998)

$$\frac{\sigma_a(t)}{\sigma_{co}} = 1 - b\ln(t^*) \tag{2.6}$$

Aydan *et al.* (2011) proposed the following function, which combines both functions above

$$\frac{\sigma_a(t)}{\sigma_{co}} = \alpha + (1 - \alpha)\frac{t^*}{1 + \beta(t^* - 1)} \tag{2.7}$$

where
α: The ultimate normalized strength of rock,
τ: The duration of short term strength (σ_{co}) test
b: empirical constant and
$t^* = \dfrac{t}{\tau}$.

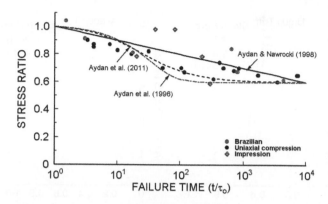

Figure 2.18 Comparison of failure time of various creep experiments and empirical relations by Aydan *et al.* (1996, 2011) and Aydan & Nawrocki (1998).

Figure 2.18 compares the failure time of samples tested in Brazilian, impression and uniaxial compression creep experiments under dry and saturated conditions. From experimental results, it is very interesting to note that if the stress ratio remains same, the failure time of dry and saturated samples are very close to each other. Furthermore, the failure times of samples tested under uniaxial compression and Brazilian creep experiments are also similar to those of impression creep experiments.

2.3 RATE-DEPENDENCY OF ROCKS AND TESTING

2.3.1 Low-rate testing of rocks

Samples obtained from Ürgüp and Avanos (Özkonak Underground City) are tested under different strain rates at the rock mechanics laboratory of Nagoya University. Some of tests results are shown in Figure 2.19. Tests on samples obtained from Derinkuyu underground city are still continuing. As seen from the figure, it seems that the strength and deformation modulus of Cappadocia tuffs decrease as the strain rate imposed on the samples decreases. These results are quite similar to those reported in rock mechanics literature (Aydan & Nawrocki, 1998). Nevertheless, further tests with greater strain rate range are felt to be necessary in order to obtain conclusive results.

2.3.2 High-rate testing of rocks

The Hopkinson Pressure Bar Testing technique was proposed by Hopkinson in 1914 as a way to measure stress pulse propagation in a metal bar (Figure 2.20). Kolsky refined Hopkinson's technique by using two Hopkinson bars in series, now known as the split-Hopkinson bar, to measure stress and strain, incorporating advancements in the cathode ray oscilloscopes in conjunction with electrical condenser units to record

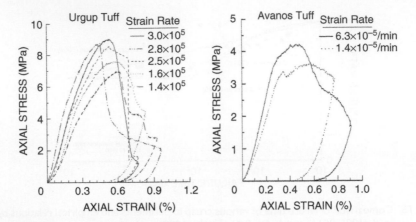

Figure 2.19 Stress-strain curves of Cappadocia tuffs under different strain rates.

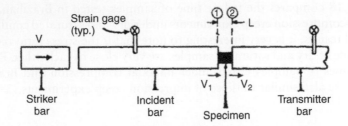

Figure 2.20 A simple illustration of split-Hopkinson Bar Technique.

the pressure wave propagation in the pressure bars. This method is used to determine dynamic properties of metals initially, and ceramics, polymers, concrete and rocks later. There are several special setups of this technique for uniaxial compression, tensile, torsion, Brazilian and triaxial compression tests of rocks. The most difficult aspect of this technique is the determination of actual straining and stresses in samples of rock.

Strains of incident bar and transmitter are measured to infer the strains and stresses of samples (Figures 2.21 and 2.22). Strains of the incident bar consist of incident strain pulse and reflected strain pulse while the strain of transmitted wave is termed as transmitted strain pulse. Stresses acting at the interfaces of the sample with incident bar and transmitter bar are given as

$$\sigma_I(t) = \frac{EA}{A_s}(\varepsilon_I(t) + \varepsilon_R(t)) \tag{2.8a}$$

$$\sigma_T(t) = \frac{EA}{A_s}(\varepsilon_T(t)) \tag{2.8b}$$

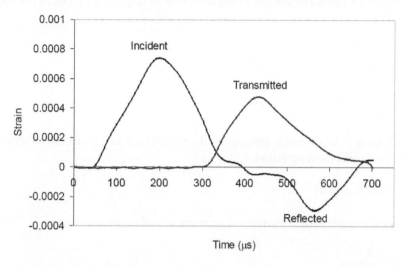

Figure 2.21 Strain pulses measured in a marble sample (from Yavuz *et al.*, 2013).

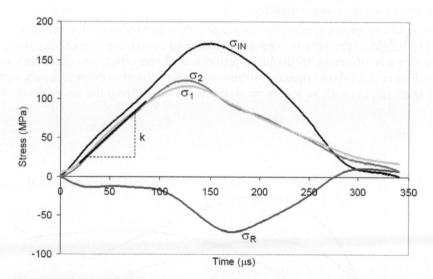

Figure 2.22 Stress pulses measured on a marble sample in a split-Hopkinson Bar experiment (from Yavuz *et al.*, 2013).

where E, A and A_s are elastic modulus and area of incident bar and area of sample. Thus, the strain rate acting on sample is inferred from the following relation

$$\dot{\varepsilon} = \frac{V_p}{L_s}(\varepsilon_I(t) - \varepsilon_R(t) - \varepsilon_T(t)) \tag{2.9}$$

Average strain and stress of sample in the direction of impact are given in the following form

$$\varepsilon(t) = \frac{V_p}{L_s} \int_0^t (\varepsilon_I(t) - \varepsilon_R(t) - \varepsilon_T(t)) dt \tag{2.10a}$$

$$\sigma(t) = \frac{EA}{2A_s} (\varepsilon_I(t) + \varepsilon_R(t) + \varepsilon_T(t)) \tag{2.10b}$$

As the following relation holds among incident, reflected and transmitted strains from the dynamic equilibrium equation

$$\varepsilon_I(t) + \varepsilon_R(t) = \varepsilon_T(t) \tag{2.11}$$

The stress on sample is directly related to transmitted strain as follow

$$\sigma(t) = \frac{EA}{A_s} \varepsilon_T(t) \tag{2.12}$$

(a) Dynamic uniaxial compression test

The uniaxial compression experiments are generally carried out using cylindrical samples. The height of the sample is generally less than its diameter. In other words, such an aspect ratio of height to diameter introduces shape effect into the experimental results. Figure 2.23 shows dynamic strain-stress relations of various rocks reported by Yavuz *et al.* (2013) with an aspect ratio of 0.6. As noted from the figure, strain-stress

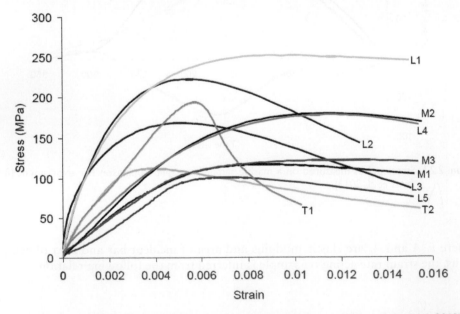

Figure 2.23 Strain-stress relations obtained in split-Hopkinson bar tests (from Yavuz *et al.*, 2013).

relations exhibit elastic-strain hardening response. Nevertheless, it should be noted such behaviour is also caused by the aspect ratio of samples.

(b) Dynamic tensile strength test (Brazilian, Notch, Slit)

Dynamic tensile strength of rocks are determined from either Brazilian Test, Semi-circular Bend tests or Cracked-Chevron Notched Brazilian Test (Figure 2.24). Tensile strength property is determined from well-known relations determined for static cases while using the stresses determined from dynamic impact. Figure 2.25 shows two example tensile strength testing of rocks reported by Cadoni *et al.* (2011)

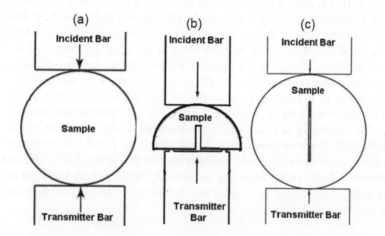

Figure 2.24 Configuration used in determination of tensile strength.

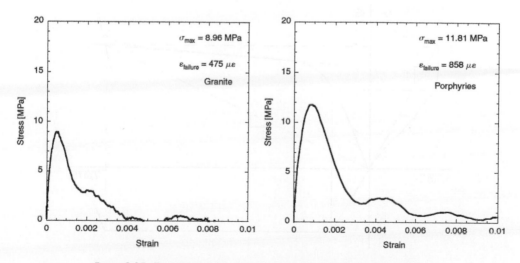

Figure 2.25 Dynamic tensile stress results (from Cadoni *et al.*, 2011).

(c) Dynamic triaxial compression test

There are some attempts to determine the dynamic strength properties under confining pressures. However, such experiments are difficult to perform and some complexities exist in the determination of stress and strains components during the experiments.

2.4   CORRELATIONS BETWEEN RATE-DEPENDENT AND CREEP TESTS

The definition of strain rate must be made in order to utilise creep test results. The strain rate in a creep test generally varies with time and it is very difficult to make such a definition. In this article, we will attempt to define it as the ratio of total strain to the time at which the tertiary creep is activated as illustrated in Figure 2.26. Furthermore, the deformation modulus or Young's modulus is defined as the ratio of the applied axial stress to the total strain at which tertiary creep is initiated as illustrated in Figure 2.6 also. Although these definitions are open to discussions, we herein attempt to re-assess both creep tests and strain rate tests in the same framework by using them.

The variation of deformation modulus or Young's modulus of Cappadocia tuffs as a function of strain rate is shown in Figure 2.27. Experimental results are highly scattered. The scattering is much more pronounced in the results of strain rate tests as compared with those of creep tests. Three functions fitted to experimental results as mean, lower and upper bounds are also shown in Figure 2.27. The deformation moduli obtained from creep tests are somewhat smaller than those obtained from strain rate tests. This might be due to the definition of strain rate employed in this section.

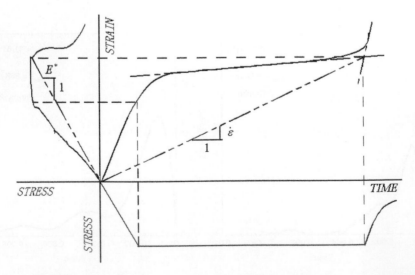

Figure 2.26 Definition of apparent strain rate and deformation modulus in a creep test response.

Next the variation of uniaxial compressive strength of Cappadocia tuff is investigated. The results obtained are shown in Figure 2.28. From this figure, it is also possible to make similar statements for the uniaxial compressive strength as in the case of the deformation modulus. Three empirical functions fitted to the experimental results as mean, lower and upper bounds are also shown in Figure 2.22.

Aydan & Nawrocki (1998) stated that the strain rate does not have any remarkable effect on the shape of yield function. This statement implies that the effect of strain rate on the yield function can be determined under uniaxial loading conditions, which simplifies the experimental procedure to a considerable extent. Taking this statement into account, the well-known Mohr-Coulomb yield criterion is presented in the following form:

$$f = \sigma_1 - (q\sigma_3 + \sigma_c(\dot{\varepsilon})) = 0 \tag{2.13}$$

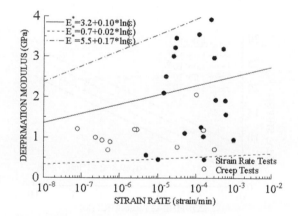

Figure 2.27 The variation of deformation modulus or Young's modulus as a function of strain rate.

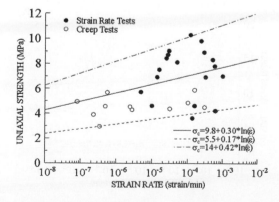

Figure 2.28 The variation of uniaxial compressive strength as a function of strain rate.

where

$$q = \frac{1 + \sin\phi}{1 - \sin\phi}$$

ϕ: Friction angle
$\sigma_c(\dot{\varepsilon})$: Uniaxial compressive strength
σ_1: Maximum principal stress
σ_3: Minimum principal stress

Using the empirical relation proposed for mean values shown in Figure 2.23, $\sigma_c(\dot{\varepsilon})$ for Cappadocia tuffs is given in the following form

$$\sigma_c(\dot{\varepsilon}) = 9.8 + 0.42\ln\dot{\varepsilon} \tag{2.14}$$

where $\dot{\varepsilon}$ is strain rate and its unit is strain/min. The unit of uniaxial compressive strength is MPa. Figures 2.29 and 2.30 shows the projections of this yield function in

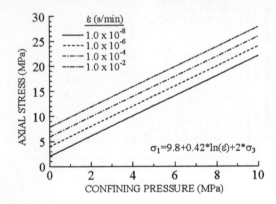

Figure 2.29 Variation of yield function as a function of confining pressure at different strain rates.

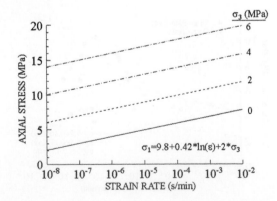

Figure 2.30 Variation of yield function as a function of strain rate at different confining pressures.

the spaces of axial stress-confining pressure and axial stress-strain rate while strain rate and confining pressure are varied, respectively. Although the yield function is simple, it is capable of taking into account the effect of the strain rate dependency of the yield function of rocks.

2.5 CONSTITUTIVE MODELING

2.5.1 Uniaxial creep laws

Most of empirical creep laws have been developed for uniaxial loading. These creep laws are generally concerned with the secondary stage, which is called the steady state creep stage. In other words, this stage should be interpreted as visco-plastic response rather than visco-elastic response as the creep rate depends upon the applied stress level.

2.5.1.1 Empirical creep laws

(a) First stage:Transient creep stage

The following creep law was proposed for the first stage or transient stage of creep behaviour:

$$\varepsilon_{ts} = Bt^{1/\beta} \tag{2.15}$$

where B and β are empirical coefficients. Andrade (1910, 1914) suggested the value of coefficient β is 3. The recent data on welded tuff by Ma & Daeman (2006) suggest that it may range between 1.8–7.8.

Lomnitz (1956, 1957) suggested the following empirical relation for the transient creep stage:

$$\varepsilon_{ts} = A \ln(1 + \alpha t) \tag{2.16}$$

The following creep function is commonly adopted in later studies, which is modification of Lomnitz law as follows:

$$\varepsilon_{ts} = A + B \log(t) + Ct \tag{2.17}$$

(b) Secondary stage: Steady state creep

A power law is used for uniaxial steady state creep strain and it is known as Norton's law

$$\varepsilon = A\sigma^n t \quad \text{or} \quad \dot{\varepsilon} = A\sigma^n \tag{2.18}$$

Typically n is about 4–5 but can range from a bit less than 2 to 8.

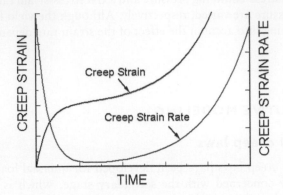

Figure 2.31 Plot of the unified creep function.

As the steady state creep does not occur below a threshold value of applied stress with respect to its short term strength, the equation above should be replaced by the following function using McCauley brackets:

$$\dot{\varepsilon} = A\left\langle \frac{\sigma_a}{\sigma_{ct}} - 1 \right\rangle^n \tag{2.19}$$

where σ_{ct} is the stress threshold to induce steady state creep response.

(c) Tertiary stage: Accelerated creep stage

Third stage creep (accelerated creep) is the final stage of the creep behaviour and it is governed by the applied boundary conditions rather than being a material property. In other words, it is a flow process. Griggs & Coles (1958) proposed the following formula for the third stage creep

$$\varepsilon_{ts} = A + Bt^2 \tag{2.20}$$

(d) Unified function for all stages

Aydan *et al.* (2003) proposed a creep function for all stages of creep response, which had the following form (Figure 2.31):

$$\varepsilon_c = A(1 - e^{-t/\tau_1}) + B(e^{t/\tau_2} - 1) \tag{2.21}$$

2.5.1.2 Simple rheological models for creep response

(a) Newton's law

Newton's law (Figure 2.32(b)) is linear and given in the following form

$$\sigma = \eta\dot{\varepsilon} \tag{2.22}$$

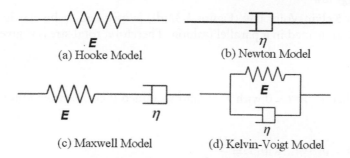

Figure 2.32 Simple rheological models.

If this law is integrated over the time, it takes the following form with a condition, that is, $\varepsilon = 0$ at $t = 0$:

$$\varepsilon = \frac{\sigma}{\eta} t \qquad (2.23)$$

If we assume that strain rate is given in the following form:

$$\dot{\varepsilon} = \frac{\sigma}{\eta} \qquad (2.24)$$

The equation above can be written as

$$\varepsilon = \dot{\varepsilon} t \qquad (2.25)$$

This has a similarity to the steady-state creep response mentioned in Section 2.4.1.

(b) Maxwell law

Substance in the Maxwell law (Figure 2.32(c)) is assumed to consist of elastic and viscous components connected in a serious fashion. Therefore, total strain and its derivative are given as

$$\varepsilon = \varepsilon_e + \varepsilon_v \quad \text{and} \quad \dot{\varepsilon} = \dot{\varepsilon}_e + \dot{\varepsilon}_v \qquad (2.26)$$

The constitutive relations for elastic and viscous responses are

$$\varepsilon_e = \frac{\sigma}{E} \quad \text{and} \quad \dot{\varepsilon}_v = \frac{\sigma}{\eta} \qquad (2.27)$$

If $\sigma = \sigma_o$ for $t \geq 0$ and $\varepsilon = \varepsilon_o$ with $\varepsilon_o = \sigma_o/E$, the above function becomes

$$\varepsilon = \frac{\sigma_o}{E} + \frac{\sigma_o}{\eta} t \qquad (2.28)$$

This equation also has a similarity to the steady-state creep response mentioned in Section 2.4.1.

(c) Kelvin-Voigt law

Substance in Kelvin-Voigt law (Figure 2.32(d)) is assumed to be elastic and viscous components connected in a parallel fashion. Therefore, total stress is given as

$$\sigma = E\varepsilon + \eta\dot{\varepsilon} \tag{2.29}$$

If stress applied σ_o at $t = 0$ with $\varepsilon = 0$ and sustained thereafter, the following relation is obtained

$$\varepsilon = \frac{\sigma_o}{E}(1 - e^{-t/t_r}) \quad \text{with } t_r = \frac{E}{\eta} \tag{2.30}$$

It is interesting to note that the above response would be similar to the transient creep stage. Figure 2.33 shows the creep strain responses for different simple rheological models.

(d) Generalized Kelvin model

The model (Figure 2.34(a)) has a Hooke element and Kelvin element connected in series fashion. The total strain of the model is

$$\varepsilon = \varepsilon_h + \varepsilon_k \tag{2.31}$$

The stress relations of each element are given as

$$\varepsilon_h = \frac{\sigma}{E_h} \quad \text{and} \quad \sigma = E_k\varepsilon_k + \eta\dot{\varepsilon}_k \tag{2.32}$$

Thus, one gets the following equation

$$\sigma = \eta\left(\dot{\varepsilon} - \frac{\dot{\sigma}}{E_h}\right) + E_k\left(\varepsilon - \frac{\sigma}{E_h}\right) \tag{2.33}$$

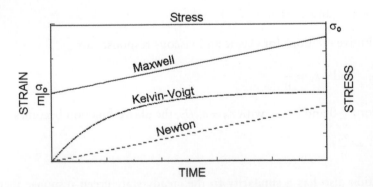

Figure 2.33 Creep strain response of simple rheological models.

If stress applied σ_o at $t = 0$ with $\varepsilon = 0$ and $\varepsilon_e = \sigma_o/E_h$ and sustained thereafter, the following relation is obtained

$$\varepsilon = \frac{\sigma_o}{E_h} + \frac{\sigma_o}{E_k}(1 - e^{-t/t_r}) \quad \text{with } t_r = \frac{\eta}{E_k} \tag{2.34}$$

As noted from this relation, instantaneous strain due to elastic response and transient creep stage can be modeled.

(e) Zener model

Zener model (Figure 2.34(b)) is also known as the standard linear solid model and it consist of a Hooke element and Maxwell element connected to each other in parallel fashion.

Total stress may be given in the following form

$$\sigma = \sigma_h + \sigma_m \tag{2.35}$$

The constitutive laws of Hooke and Maxwell elements are

$$\sigma_h = E_h\varepsilon; \quad \varepsilon = \varepsilon_s + \varepsilon_d; \quad \dot{\varepsilon} = \dot{\varepsilon}_s + \dot{\varepsilon}_d; \quad \dot{\varepsilon}_s = \frac{\dot{\sigma}_m}{E_m}; \quad \dot{\varepsilon}_d = \frac{\sigma_m}{\eta_m} \tag{2.36}$$

Thus, one can easily get the following differential equation

$$\frac{d\varepsilon}{dt} + \frac{1}{\eta_m} \cdot \frac{E_h E_m}{E_h + E_m}\varepsilon = \frac{1}{E_h + E_m}\left(\frac{d\sigma}{dt} + \frac{E_m}{\eta_m}\sigma\right) \tag{2.37}$$

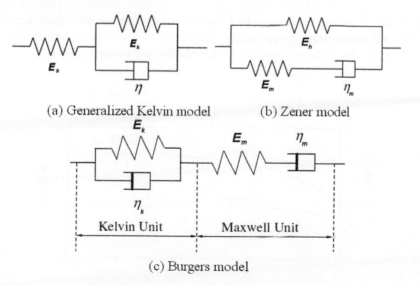

(a) Generalized Kelvin model   (b) Zener model

Kelvin Unit   Maxwell Unit

(c) Burgers model

Figure 2.34 More complex rheological models.

If stress applied σ_o at $t = 0$ with and $\varepsilon_o = \sigma_o/(E_h + E_m)$ and sustained thereafter, the following relation is obtained

$$\varepsilon = \frac{\sigma_o}{E_h}\left[1 - \frac{E_m}{E_h + E_m}e^{-t/t_r}\right] \quad \text{with } t_r = \eta_m\frac{E_m + E_h}{E_m E_h} \tag{2.38}$$

The creep response to be determined from this model involves the instantaneous strain and transient creep.

(f) Burgers model

Burgers model (Figure 2.34(c)) consists of Maxwell and Kelvin elements connected to each other in a series fashion. The constitutive relations for each element can be given as

$$\dot{\varepsilon}_m = \frac{\dot{\sigma}}{E_m} + \frac{\sigma}{\eta_m} \quad \text{and} \quad \sigma = E_k\varepsilon_k + \eta\dot{\varepsilon}_k \tag{2.39}$$

The total strain is given by

$$\varepsilon = \varepsilon_m + \varepsilon_k \tag{2.40}$$

If stress applied σ_o at $t = 0$ with $\varepsilon = 0$ and $\varepsilon_m = \sigma_o/E_m$ and sustained thereafter, the following relation is obtained

$$\varepsilon = \frac{\sigma_o}{E_m} + \frac{\sigma_o}{E_k}(1 - e^{-t/t_k}) + \frac{\sigma_o}{\eta_m}t \quad \text{with } t_k = \frac{\eta_k}{E_k} \tag{2.41}$$

As noted, this model can simulate the instantaneous strain due to elastic response and transient and steady state creep stages. Figure 2.35 shows and compares the creep strain responses for different more complex rheological models.

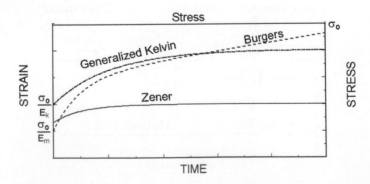

Figure 2.35 Creep responses from more complex rheological models.

(g) Visco-plastic models

(1) Bingham model – elastic-perfectly visco-plastic model

Visco-plastic model of Bingham type assumes that the material behaves elastically below the yield stress level and visco-plastic above the yield stress level given as:

$$\varepsilon = \frac{\sigma}{E} \quad \text{if } \sigma < \sigma_o \tag{2.42}$$

$$\varepsilon = \frac{\sigma - \sigma_o}{\eta} t + \frac{\sigma}{E} \quad \text{if } \sigma > \sigma_o \tag{2.43}$$

The equation above corresponds to the perfectly visco-plastic material if σ_o corresponds to yield threshold value of stress. Furthermore, the fluidity coefficient is defined as

$$\gamma = \frac{1}{\eta} \tag{2.44}$$

(2) Elastic-visco-plastic model of hardening type (Prezyna type)

Elastic-visco-plastic model of hardening type (Prezyna type) (Figure 2.36) assumes that the material behaves elastically below the yield stress level and visco-plastic above the yield stress level σ_Y given as: The yield strength of visco-plastic material in relation to the visco-plastic strain of hardening type can be written as

$$Y = \sigma_Y + H\varepsilon_{vp} \tag{2.45}$$

Furthermore, total strain is assumed to be a sum of elastic strain and visco-plastic strain as

$$\varepsilon = \varepsilon_e + \varepsilon_{vp} \tag{2.46}$$

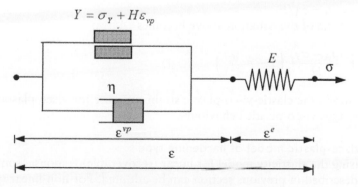

Figure 2.36 Elastic-visco-plastic model.

Thus the stress-strain relations are given in the following form

$$\sigma_p = \sigma = E\varepsilon \quad \text{if } \sigma_p < Y \tag{2.47}$$

$$\sigma_p = \sigma_Y + H\varepsilon_{vp} \quad \text{if } \sigma_p > Y \tag{2.48}$$

Total stress at any time can be written as

$$\sigma = \sigma_p + \sigma_d \tag{2.49}$$

Viscous component of stress is related to visco-plastic strain rate as follows

$$\sigma_d = C_p \frac{d\varepsilon_{vp}}{dt} \tag{2.50}$$

Thus, one can obtain the following differential equation for visco-plastic response

$$\sigma = \sigma_Y + H\varepsilon_{vp} + C_p \frac{d\varepsilon_{vp}}{dt} \tag{2.51}$$

Replacing the visco-plastic strain with the use of total strain and elastic strain in the above equation, one can easily obtain the following relation

$$HE\varepsilon + \frac{1}{C_p}E\frac{d\varepsilon}{dt} = H\sigma + E(\sigma - \sigma_Y) + \frac{1}{C_p}\frac{d\sigma}{dt} \tag{2.52}$$

Let us assume that a constant stress $\sigma = \sigma_A$ is applied and kept constant (creep test). The differential equation above is reduced to the following form:

$$\frac{d\varepsilon}{dt} + \frac{H}{E}\varepsilon = \frac{H}{C_pE}\sigma_A + \frac{1}{C_p}(\sigma_A - \sigma_Y) \tag{2.53}$$

The solution of the differential equation above is obtained as follows

$$\varepsilon = Ce^{-\frac{H}{C_p}t} + \frac{1}{E}\sigma_A + \frac{1}{H}(\sigma_A - \sigma_Y) \tag{2.54}$$

when $t = 0$, $\varepsilon = \varepsilon_e = \sigma_A/E$.
The final form of the equation above becomes

$$\varepsilon = \frac{\sigma_A}{E} + \frac{(\sigma_A - \sigma_Y)}{H}\left(1 - e^{-\frac{H}{C_p}t}\right) \tag{2.55}$$

Figure 2.37 shows the elastic-visco-plastic strain response for visco-plastic hardening and Bingham type visco-plastic behaviours.

(3) Elasto-visco-plastic model of hardening type
Instead of using the elasticity model for linear (recoverable) response, some rheological models described in previous section can be adopted. For non-linear (permanent) response, the models described can be utilized. For example, if the linear response

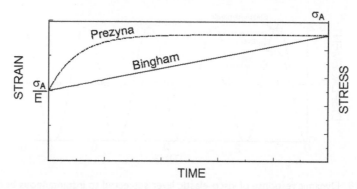

Figure 2.37 Responses obtained from elastic-visco-plastic models.

is modeled using the Kelvin-Voigt type model, the following relation would hold for linear part $(\sigma < \sigma_y)$

$$\varepsilon_r = \frac{\sigma}{E} \quad \text{and} \quad \sigma = E\varepsilon_r + \eta\dot{\varepsilon}_r \quad \text{with } \varepsilon = \varepsilon_r \tag{2.56}$$

As for the non-linear (permanent) part $\sigma \geq \sigma_y$, the following can be written

$$\sigma = \sigma_Y + H\varepsilon_p + C_p \frac{d\varepsilon_p}{dt} \tag{2.57}$$

Total strain is assumed to consist of linear (recoverable) and non-linear (permanent) components as given below

$$\varepsilon = \varepsilon_r + \varepsilon_p \tag{2.58}$$

(h) Dynamic creep responses of Kelvin-Voigt type

Aydan (1994, 1997) presented a dynamic creep response of rocks or layers with the use of Kelvin-Voigt model. Most of creep functions assume that the elastic response takes instantaneously. Instead Aydan considered the dynamic equilibrium equations and he suggested that the short–term dynamic loading response should yield the linear viscosity characteristics of rocks.

(1) Visco-elastic layer subjected to instantaneous body force

The original derivation can be adopted and the final form of displacement response at the top of the layer with a thickness L would take the following form for complex roots of the characteristics of differential equation:

$$u(t) = \frac{\rho g}{L^2} \cdot \frac{1}{p^2 + q^2} \left[1 - e^{-pt} \left(\cos qt + \frac{p}{q} \sin qt \right) \right] \tag{2.59}$$

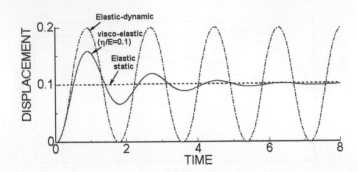

Figure 2.38 Dynamic response of visco-elastic layer subjected to instantaneous body force.

where

$$p = \frac{3\eta}{2\rho g L^2} \quad \text{and} \quad q = \frac{1}{\rho g L^2}\sqrt{12E\rho L^2 - 9\eta^2}$$

Figure 2.38 shows the dynamic and static responses of the layer subjected to instantaneous body force.

(2) Uniaxial compression creep experiment
The details of the derivation of equations for strain-stress responses during a uniaxial compression experiments are presented by Aydan (1997). The final forms of equations for responses take the following forms for complex roots of the characteristics of the differential equation:

(i)   Linear load increase stage $\left(0 \le t \le T_o; \; \sigma = \sigma_o \frac{t}{T_o}\right)$

$$\varepsilon(t) = e^{-pt}(C_1 \cos qt + C_2 \sin qt) + \frac{\sigma_o}{\rho L^2 T_o} \cdot \frac{1}{(p^2 + q^2)^2}[(p^2 + q^2)t + 2p] \quad (2.60)$$

(ii)   Load constant stage $(t \ge T_o; \; \sigma = \sigma_o)$

$$\varepsilon(t) = e^{-pt}(C_1 \cos qt + C_2 \sin qt) + \frac{\sigma_o}{\rho L^2 T_o} \cdot \frac{1}{(p^2 + q^2)} \quad (2.61)$$

where

$$p = \frac{\eta}{2\rho L^2} \quad \text{and} \quad q = \frac{1}{2\rho L^2}\sqrt{4E\rho L^2 - \eta^2}$$

L: height of sample; ρ: density; E: elastic modulus; η: viscosity modulus

Figure 2.39 shows the stress-strain responses of a sample under uniaxial compression condition. In the computations, the viscosity coefficient was varied. As noted

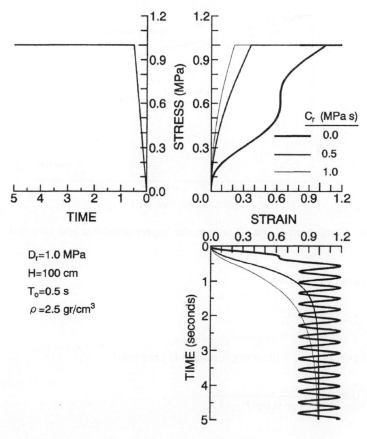

Figure 2.39 Dynamic visco-elastic response during uniaxial creep experiments.

from the figure, when rock is purely elastic, the strain response would oscillate in time. Therefore, the short term-dynamic response may be useful for characterizing the transient viscous properties of rocks, particularly.

(3) Long term strength
The strength of rocks is assumed to be hardening type in the approaches described above. However, it is well known that the long term strength ($\sigma_a(t)$) of rocks decreases with time and it is expressed in the following forms and they are compared in Figure 2.40 for experimental data on Oya tuff:

Aydan *et al.* (1996)

$$\frac{\sigma_a(t)}{\sigma_{co}} = 0.6 + 0.3e^{-0.03(t^*-1)} \tag{2.62}$$

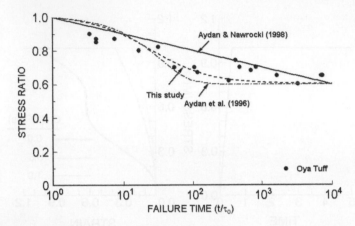

Figure 2.40 Comparison of uniaxial compression experimental data with empirical relations.

Aydan & Nawrocki (1998)

$$\frac{\sigma_a(t)}{\sigma_{co}} = 1 - 0.1\ln(t^*) \qquad (2.63)$$

In this note another, the following function is proposed

$$\frac{\sigma_a(t)}{\sigma_{co}} = 0.6 + 0.4\frac{t^*}{1 + 0.04(t^* - 1)} \qquad (2.64)$$

where
α: The ultimate normalized strength of rock
τ: The duration of short term strength (σ_{co}) test
b: empirical constant and
$t^* = \dfrac{t}{\tau}$.

2.5.2 Multi-dimensional constitutive laws

2.5.2.1 Linear constitutive laws

When rock or rock mass behaves linearly without any rate dependency, the simplest constitutive law is Hooke's law. This law is written in the following form:

$$\sigma_{ij} = D_{ijkl}\varepsilon_{kl} \qquad (2.65)$$

where σ_{ij}, ε_{kl} and D_{ijkl} are stress, strain and elasticity tensors, respectively.

If material is homogenous and isotropic, Eq. (2.65) may be written as

$$\sigma_{ij} = 2\mu\varepsilon_{ij} + \lambda\delta_{ij}\varepsilon_{kk} \qquad (2.66)$$

where δ_{ij} is Kronecker delta tensor. λ and μ are Lame coefficients, which are given in terms of elasticity (Young's) modulus (E) and Poisson's ratio (v) as

$$\lambda = \frac{Ev}{(1+v)(1-2v)}; \quad \mu = \frac{E}{2(1+v)} \tag{2.67}$$

When rock or rock mass behaves linearly with rate dependency, which may be called visco-elasticity. One of simplest constitutive laws is Voigt-Kelvin Law, which may written in the following form:

$$\sigma_{ij} = D_{ijkl}\varepsilon_{kl} + C_{ijkl}\dot{\varepsilon}_{kl} \tag{2.68}$$

where $\dot{\varepsilon}_{kl}$ and C_{ijkl} are strain rate and viscosity tensors, respectively.

If material is homogenous and isotropic, Eq. (2.68) may be written in analogy to Eq. (2.67) according to Aydan (1995) as

$$\sigma_{ij} = 2\mu\varepsilon_{ij} + \lambda\delta_{ij}\varepsilon_{kk} + 2\mu^*\dot{\varepsilon}_{ij} + \lambda^*\delta_{ij}\dot{\varepsilon}_{kk} \tag{2.69}$$

Coefficients λ^* and μ^* may be called viscous Lame coefficients. There are different visco-elasticity models as described in the previous section. The above constitutive laws have to be replaced with their equivalents.

2.5.2.2 Non-linear behaviour (elasto-plasticity and elasto-visco-plasticity)

(a) Elasto-plasticity

Every material in nature starts to yield after a certain stress or strain state and rock or rock mass is a no exception. The terms used to describe the material behaviour such as elasticity and visco-elasticity are replaced by the terms of elasto-plasticity or elasto-visco-plasticity as soon as material behaviour deviates from linearity. The relation between total stress and strain or strain rate can no longer be used and every relation must be written in incremental form. For example, if the conventional plasticity models were used, the elasto-plastic constitutive law between incremental stress and strain tensors would take the following form:

$$\Delta\sigma_{ij} = D_{ijkl}^{ep}\Delta\varepsilon_{kl} \tag{2.70}$$

where

$$D_{ijkl}^{ep} = \left(D_{ijkl} - \frac{D_{ijmn}\dfrac{\partial F}{\partial \sigma_{mn}}\dfrac{\partial G}{\partial \sigma_{pr}}D_{prkl}}{h + \dfrac{\partial F}{\partial \sigma_{mn}}D_{mnpr}\dfrac{\partial G}{\partial \sigma_{pr}}} \right) \tag{2.71}$$

The specific derivation of Eqs. (2.70) and (2.71) requires the followings

✓   Existence of a yield function (Mohr-Coulomb, Drucker-Prager etc.),
✓   Flow rule (existence of a plastic potential function),

✓ Prager's consistency condition,
✓ Linear decomposition of incremental strain tensor into elastic and plastic components, and
✓ Existence of Hooke's law between incremental stress and elastic strains.

The plastic strain of metals is generally assumed to be independent of the volumetric response and it is quite common to introduce the effective stress (σ_e) and effective strain (ε_e) concepts as given below

$$\sigma_e = \sqrt{\frac{3}{2} \mathbf{s} \cdot \mathbf{s}} \quad \text{and} \quad \varepsilon_e = \sqrt{\mathbf{e}_p \cdot \mathbf{e}_p} \tag{2.72}$$

where $\mathbf{s}$ and $\mathbf{e}$ are deviatoric stress and deviatoric strain tensors are given as follow

$$\mathbf{s} = \boldsymbol{\sigma} - \frac{\text{tr}(\boldsymbol{\sigma})}{3}\mathbf{I} \quad \text{and} \quad \mathbf{e}_p = \boldsymbol{\varepsilon}_p - \frac{\text{tr}(\boldsymbol{\varepsilon}_p)}{3}\mathbf{I} \quad \text{with } \text{tr}(\boldsymbol{\varepsilon}_p) = 0 \tag{2.73}$$

It is interesting to note that the effective stress and strain would corresponds to those at uniaxial state, that is,

$$\sigma_e = \sigma_1 \quad \text{and} \quad \varepsilon_e = \varepsilon_1 \tag{2.74}$$

This is a very convenient conclusion that the non-linear response can be evaluated under uniaxial state and can be easily extended to multi-dimensional state without any triaxial testing. However, it should be noted that this is only valid when the volumetric components are negligible in the overall mechanical behaviour.

(b) Elastic-visco-plasticity

These approaches assume that the materials are assumed to be elastic before yielding and behave in a visco-plastic manner following yielding. In visco-plastic evaluations, $\mathbf{e}_p$ is replaced by $\mathbf{e}_{vp}$.

(i) Power-type Model
   When Norton type constitutive law is used for creep response, the visco-plastic strain rate ($\mathbf{e}_{vp} = \boldsymbol{\varepsilon}_{vp}$) is expressed as follows:

$$\frac{d\boldsymbol{\varepsilon}_{vp}}{dt} = \left(\frac{\sigma_{eq}}{\sigma_o}\right)^n \frac{\partial \sigma_{eq}}{\partial \boldsymbol{\sigma}} \tag{2.75}$$

(ii) Perzyna-type
   Perznya-type elastic-visco-plastic laws are used for representing non-linear rate dependency involving plasticity.

$$\frac{d\boldsymbol{\varepsilon}_{vp}}{dt} = \lambda \mathbf{s} \tag{2.76}$$

where λ is proportionality coefficient and it is interpreted as fluidity coefficient. This parameter is obtained from uniaxial creep experiments as

$$\lambda = \frac{\dot{\varepsilon}_c}{\sigma} \tag{2.77}$$

(d) Elasto-visco-plasticity

Another approach was proposed by Aydan & Nawrocki (1998), in which the material behaviour is visco-elastic before yielding and becomes visco-plastic after yielding. The derivation of this constitutive law involves the followings:

1) Yield function

$$F(\sigma, \kappa_p, \kappa_v) = f(\sigma) - K(\kappa_p, \kappa_v) = 0 \tag{2.78}$$

It should be noted that the yield function is a function of permanent plastic and viscous hardening parameters (Figure 3.9).

2) Flow rule

$$d\varepsilon^p = \lambda \frac{\partial G}{\partial \sigma}, \quad d\dot{\varepsilon}^p = \dot{\lambda} \frac{\partial G}{\partial \sigma} + \lambda \frac{\partial \dot{G}}{\partial \sigma} \tag{2.79}$$

3) Prager's consistency condition

$$dF = \frac{\partial F}{\partial \sigma} \cdot d\sigma + \frac{\partial F}{\partial \kappa_p} \frac{\partial \kappa_p}{\partial \varepsilon_p} \cdot d\varepsilon_p + \frac{\partial F}{\partial \kappa_v} \frac{\partial \kappa_v}{\partial \dot{\varepsilon}_p} \cdot d\dot{\varepsilon}_p = 0 \tag{2.80}$$

4) Linear decomposition of the strain increment ($d\varepsilon$) and strain rate increment ($d\dot{\varepsilon}$) into their reversible ($d\varepsilon^r$) and permanent components ($d\varepsilon^p$)

$$d\varepsilon = d\varepsilon^r + d\varepsilon^p; \quad d\dot{\varepsilon} = d\dot{\varepsilon}^r + d\dot{\varepsilon}^p \tag{2.81}$$

5) Incremental Kelvin-Voigt law

$$d\sigma = \mathbf{D}^r d\varepsilon^r + \mathbf{C}^r d\dot{\varepsilon}^r \tag{2.82}$$

where
σ: stress tensor
ε: strain tensor
$K(\kappa_p, \kappa_v)$: hardening function
G: plastic potential
λ: proportionality coefficient
κ_p: plastic hardening parameter
κ_v: viscous hardening parameter
$d\varepsilon^r$: Reversible incremental strain tensor
$d\dot{\varepsilon}^r$: Reversible incremental strain rate tensor
$d\varepsilon^p$: Permanent incremental strain tensor
$d\dot{\varepsilon}^p$: Permanent incremental strain rate tensor

$\mathbf{D}^r$: Elasticity tensor
$\mathbf{C}^r$: Viscosity tensor
$(\cdot)$ denotes dot product

In elastic-visco-plastic formulations of Perzyna type, the flow rule is assumed to be of the following form

$$d\dot{\varepsilon}^p = \dot{\lambda}\frac{\partial G}{\partial \sigma} \tag{2.83}$$

The flow rule above implies that any plastic straining is time-dependent. Aydan & Nawrocki (1998) suggested the following form:

$$d\dot{\varepsilon}^p = \lambda\frac{\partial \dot{G}}{\partial \sigma} \tag{2.84}$$

This flow rule of Aydan & Nawrocki (1998) implies that the plastic potential function shrinks (or expand) in time domain while keeping its original form in stress space and the permanent strain increment consists of time-dependent and time independent parts.

Substituting Eq. (2.84) in Eq. (2.80) and re-arranging the resulting equations yields the following

$$dF = \frac{1}{h_{rp}}\frac{\partial F}{\partial \sigma} \cdot d\sigma \tag{2.85}$$

where h_{rp} is called hardening modulus and given specifically as follows

$$h_{rp} = -\left[\frac{\partial F}{\partial \kappa_p}\frac{\partial \kappa_p}{\partial \varepsilon_p} \cdot \frac{\partial G}{\partial \sigma} + \frac{\partial F}{\partial \kappa_v}\frac{\partial \kappa_v}{\partial \dot{\varepsilon}_p} \cdot \frac{\partial \dot{G}}{\partial \sigma}\right] \tag{2.86}$$

Inserting the relations above into Eq. (2.80) and (2.81) yields the constitutive relations between permanent strain increment and permanent strain rate increment and stress increment as

$$d\varepsilon_p = \frac{1}{h_{rp}}\frac{\partial G}{\partial \sigma}\left(\frac{\partial F}{\partial \sigma} \cdot d\sigma\right) = \frac{1}{h_{rp}}\left(\frac{\partial G}{\partial \sigma} \otimes \frac{\partial F}{\partial \sigma}\right) \cdot d\sigma \tag{2.87}$$

$$d\dot{\varepsilon}_p = \frac{1}{h_{rp}}\frac{\partial \dot{G}}{\partial \sigma}\left(\frac{\partial F}{\partial \sigma} \cdot d\sigma\right) = \frac{1}{h_{rp}}\left(\frac{\partial \dot{G}}{\partial \sigma} \otimes \frac{\partial F}{\partial \sigma}\right) \cdot d\sigma \tag{2.88}$$

where $(\otimes)$ denotes the tensor product. The inverse of the relations above cannot be determined whether the plastic potential function is of associated or non-associated type. Therefore the following technique is used to establish the relation between stress increment and strain and strain rate increments. Using relations given by (2.79), (2.80), (2.87) and (2.88), one can write the followings:

$$d\sigma = \mathbf{D}^r d\varepsilon - \mathbf{D}^r \frac{1}{h_{rp}} \frac{\partial G}{\partial \sigma} \left(\frac{\partial F}{\partial \sigma} \cdot d\sigma \right) + \mathbf{C}^r d\dot{\varepsilon} - \mathbf{C}^r \frac{1}{h_{rp}} \frac{\partial \dot{G}}{\partial \sigma} \left(\frac{\partial F}{\partial \sigma} \cdot d\sigma \right) \tag{2.89}$$

Taking the dot products of the both sides of the expression above by $\partial F / \partial \sigma$ yields

$$\frac{\partial F}{\partial \sigma} \cdot d\sigma = \frac{\dfrac{\partial F}{\partial \sigma} \cdot (\mathbf{D}^r d\varepsilon) + \dfrac{\partial F}{\partial \sigma} \cdot (\mathbf{C}^r d\dot{\varepsilon})}{1 + \dfrac{1}{h_{rp}} \dfrac{\partial F}{\partial \sigma} \cdot \left(\mathbf{D}^r \dfrac{\partial G}{\partial \sigma} \right) + \dfrac{1}{h_{rp}} \dfrac{\partial F}{\partial \sigma} \cdot \left(\mathbf{C}^r \dfrac{\partial \dot{G}}{\partial \sigma} \right)} \tag{2.90}$$

Substituting the equation above in Eq. (2.82) gives the incremental elasto-visco-plastic constitutive law as

$$d\sigma = \mathbf{D}^{rp} d\varepsilon + \mathbf{C}^{rp} d\dot{\varepsilon} \tag{2.91}$$

where

$$\mathbf{D}^{rp} = \mathbf{D}^r - \frac{\mathbf{D}^r \dfrac{\partial G}{\partial \sigma} \otimes \dfrac{\partial F}{\partial \sigma} \mathbf{D}^r}{h_{rp} + \dfrac{\partial F}{\partial \sigma} \cdot \left(\mathbf{D}^r \dfrac{\partial G}{\partial \sigma} \right) + \dfrac{\partial F}{\partial \sigma} \cdot \left(\mathbf{C}^r \dfrac{\partial \dot{G}}{\partial \sigma} \right)} \tag{2.92}$$

$$\mathbf{C}^{rp} = \mathbf{C}^r - \frac{\mathbf{C}^r \dfrac{\partial \dot{G}}{\partial \sigma} \otimes \dfrac{\partial F}{\partial \sigma} \mathbf{C}^r}{h_{rp} + \dfrac{\partial F}{\partial \sigma} \cdot \left(\mathbf{D}^r \dfrac{\partial G}{\partial \sigma} \right) + \dfrac{\partial F}{\partial \sigma} \cdot \left(\mathbf{C}^r \dfrac{\partial \dot{G}}{\partial \sigma} \right)} \tag{2.93}$$

Figure 2.41 illustrates the elasto-visco-plastic model for one-dimensional response. Non-linear behaviour requires the existence of yield functions. These yield functions are also called failure functions at the ultimate state when rocks rupture. For two-dimensional case, it is common to use the Mohr-Coulomb yield criterion given by:

$$\tau = c + \sigma_n \tan \phi \quad \text{or} \quad \sigma_1 = \sigma_c + q\sigma_3 \tag{2.94}$$

where c, ϕ and σ_c are cohesion, friction angle and uniaxial compressive strength. σ_c and q are related to cohesion and friction angle in the following form

$$\sigma_c = \frac{2c \cos \phi}{1 - \sin \phi} \quad \text{and} \quad q = \frac{1 + \sin \phi}{1 - \sin \phi} \tag{2.95}$$

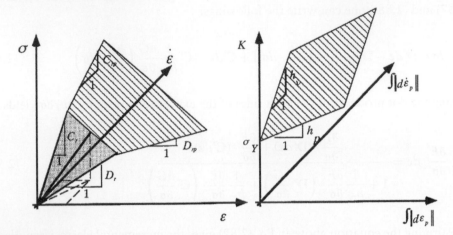

Figure 2.41 The elasto-visco-plastic model for one-dimensional response.

Since the intermediate principal stress is indeterminate in Mohr-Coulomb criterion and there is a corner-effect problem during the determination of incremental elasto-plasticity tensor, the use of Drucker-Prager criterion is quite common in numerical analyses, which is given by

$$\alpha I_1 + \sqrt{J_2} = k \tag{2.96}$$

where

$$I_1 = \sigma_1 + \sigma_2 + \sigma_3; \quad J_2 = \frac{1}{6}((\sigma_1 - \sigma_2)^2 + (\sigma_2 - \sigma_3)^2 + (\sigma_3 - \sigma_1)^2$$

Nevertheless, it is possible to relate the Drucker-Prager yield criterion with the Mohr-Coulomb yield criterion. On π-plane, if the inner corners of the Mohr-Coulomb yield surface are assumed to coincide the Drucker-Prager yield criterion, the following relations may be derived

$$\alpha = \frac{2\sin\phi}{\sqrt{3}(3 + \sin\phi)}; \quad k = \frac{6c\cos\phi}{\sqrt{3}(3 + \sin\phi)} \tag{2.97}$$

In Rock Mechanics, one of recent yield criterion is Hoek-Brown's criterion (1980), which is written as

$$\sigma_1 = \sigma_3 + \sqrt{m\sigma_c\sigma_3 + s\sigma_c^2} \tag{2.98}$$

where m and s are some coefficients. While the value of s is 1 for intact rock, the values of m and s change when they are used for rock mass. The value of m can be

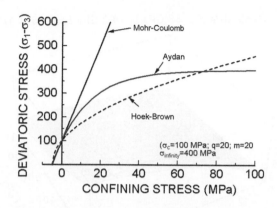

Figure 2.42 Comparison of various yield functions (from Aydan, 2008).

related to tensile strength and uniaxial compressive strength of rock requiring $\sigma_1 = 0$ and $\sigma_3 = -\sigma_t$ as

$$m = \frac{\sigma_c^2 - \sigma_t^2}{\sigma_c \sigma_t} \tag{2.99}$$

It must be noted this yield criterion cannot be applied to non-cohesive frictional materials, that is $\sigma_c = 0$ or $c = 0$.

Aydan (1995) proposed a yield function for thermo-plasticity yielding of rock as given by

$$\sigma_1 = \sigma_3 + [S_\infty - (S_\infty - \sigma_c)e^{-b_1\sigma_3}]e^{-b_2 T} \tag{2.100}$$

where S_∞ is ultimate deviatoric strength while coefficients b_1, b_2 are empirical constants. Figure 2.42 compares the yield functions of Mohr-Coulomb, Hoek and Brown and Aydan and Figure 2.43 shows the yield function of Aydan in the space of confining pressure and temperature.

The parameters of yield function given above needs to be modified for time-dependent behaviour as illustrated in Figure 2.44. As pointed by Aydan *et al.* (1994, 2010) and Aydan & Nawrocki (1998), the experimental results indicate that the time-dependency of the friction angle of rocks is quite negligible. Therefore, the parameters related to the cohesion of rocks may only be necessary to be correlated with time-dependency. Empirical proposals given by Eq. (2.5) to (2.7) may be used for the time-dependency of cohesion-related parameters. The function below is such an example and was used by Aydan *et al.* (2010) for assessing the long term response of an underground power house:

$$\frac{c(t)}{c_o} = 1.0 - 0.0282 \log(t^*) \tag{2.101}$$

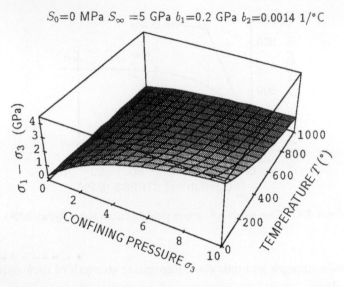

$S_0 = 0$ MPa $S_\infty = 5$ GPa $b_1 = 0.2$ GPa $b_2 = 0.0014$ 1/°C

Figure 2.43 3D view of Aydan's yield function (from Aydan, 1995).

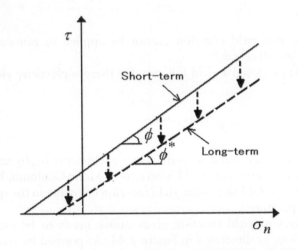

Figure 2.44 An illustration of long term variation of Mohr-Coulomb yield criterion.

where

c_o: cohesion obtained from short term experiment with a duration t_o,

$$t^* = \frac{t}{t_o},$$

t: time.

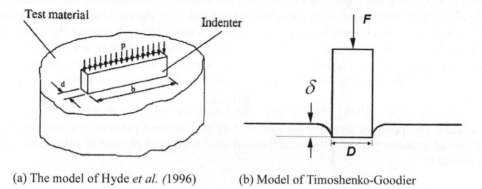

(a) The model of Hyde *et al.* *(1996)*   (b) Model of Timoshenko-Goodier

Figure 2.45 Geometrical illustration of various models.

2.6 CORRELATION BETWEEN COMPRESSION CREEP TESTS AND IMPRESSION CREEP TESTS

Impression creep experiments are relatively easy to perform and the capacity of loading equipment is relatively small compared to conventional creep experiments. Various approaches on the relation between stress and strain states of impression experiments and conventional compression experiments are summarized and discussed in this section.

2.6.1 Empirical correlations

Hyde *et al.* (1996) suggested the following empirical relations between the strain measured in conventional creep tests and indenter with a rectangular flat end with the use of finite element studies (Figure 2.45(a)):

$$\sigma = \eta p \quad \text{and} \quad \varepsilon = \frac{\delta}{\beta d} \tag{2.102}$$

They analysed various configurations using finite element technique and concluded that η and β range between 0.43–0.47 and 1.9–2.0 for a stripe-like flat-end indenters, respectively. Sun & Hyde (2009) suggests 0.892 and 0.448 for the values of η and β for a circular flat-end indenter.

2.6.2 Analytical correlations

The stress and strain field induced in the impression experiments is close to the compression of the rock under a rigid indenter (Figure 4.1(b)). Timoshenko & Goodier (1951) developed the following relation for a circular rigid indentation of elastic half-space problem:

$$\frac{\delta}{D} = \frac{\pi}{4} \frac{1 - \nu^2}{E} p \quad \text{with } p = \frac{4F}{\pi D^2} \tag{2.103}$$

where F, v and E are applied load, Poisson ratio and elastic modulus of rock. While the displacement distribution is uniform beneath the indenter, the contact pressure induced by the indenter would not be uniform. Jaeger & Cook (1969) discussed the initiation of yielding due to compression and they suggested that the yielding stress level under compression should correspond to one to two times the uniaxial compressive strength of rock and the yielding would occur at a depth of the order of the radius of the indenter.

Aydan *et al.* (2008) showed that the following relation should exist between uniaxial compression stress-strain rate ($\sigma - \varepsilon$) and applied pressure (p) and nominal strain of indenter with a cylindrical flat-end ($\varepsilon_i = \delta/D$) with the use of spherical cavity approach:

$$\frac{\sigma}{\varepsilon} = \frac{1+v}{2}\frac{p}{\varepsilon_i} \tag{2.104}$$

where v is Poisson's ratio of rock and D is the diameter of the indenter. It should be noted that radial stress and strain of impression experiments are analogous to stress-strain of uniaxial stress state. As discussed by Aydan *et al.* (2008), there may be at least two stress levels for initiating the yielding of rock beneath the indenter. The first yield stress level would correspond to twice the tensile strength level of rock and the other one would correspond to uniaxial compressive strength level. However, the effect of tensile yielding is generally difficult to differentiate as the deformation moduli before and after yielding in tension remains fairly the same. The overall deformation modulus may change after yielding at in compression. The experiments also indicate that the ultimate strength value (p_u) cannot be greater that a stress level given by

$$p_u = \frac{2}{1 - \sin\phi}\sigma_c \tag{2.105}$$

The equation above implies that the ultimate strength for a frictionless cohesive medium would be twice its uniaxial strength or four times its cohesion. However, it should be noted that this type equation implies that considerable yielding should take place beneath the indenters. Aydan *et al.* (2008) developed following formulas for three different situations of rock beneath the indenter (Figure 2.46):

Elastic behaviour ($p_i \leq 2\sigma_t$)

$$\frac{u_a}{a} = \frac{1+v}{2E}p_i \tag{2.106}$$

Radially ruptured (no tension) plastic behaviour ($2\sigma_t < p_i \leq \sigma_c$)

$$\frac{u_a}{a} = \frac{1+v}{2E}p_t\frac{R_t}{a} + \frac{p_i}{2E}\left(1 - \frac{a}{R_t}\right); \quad \frac{R_t}{a} = \left(\frac{p_i}{p_t}\right)^{1/2}; \quad p_t \leq 2\sigma_t \tag{2.107}$$

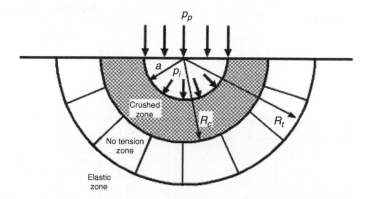

Figure 2.46 An illustration of zones formed beneath loading plate and notation.

Crushed Plastic Behaviour ($p_i > \sigma_c$)

$$\frac{u_a}{a} = \left[\frac{1+\nu}{2E} p_t \frac{R_t}{a} + \frac{p_c}{2E} \frac{R_c}{a}\left(1 - \frac{R_c}{R_t}\right)\right]\left(\frac{p_i}{p_c}\right)^{q/(q-1)}; \quad \frac{R_c}{a} = \left(\frac{p_i}{p_c}\right)^{q/2(q-1)}; \quad \frac{R_c}{R_t} = \left(\frac{p_t}{p_c}\right)^{1/2}$$

(2.108)

Furthermore, the applied pressure is equal to radial pressure on the walls of spherical body in view of the equivalence of work done by the pressure of the indenter to that induced by the wall of spherical body on the surrounding medium as

$$p_p = p_i$$

(2.109)

Assuming that the volume of hemispherical body beneath the indenter remains for a given impression displacement (δ), the outward displacement (u_a) of the hemispherical cavity wall can be easily related to the impression displacement (δ) as follows

$$\delta = 2u_a$$

(2.110)

2.6.3 Numerical studies on correlations between experimental techniques

Finite element method is best suited for studying stress-strain responses of various objects under different loading regimes. An axisymmetric finite element analysis of cylindrical indenter with a diameter of 3 mm under elastic behaviour was carried out (Figure 2.47(a)). The material properties on the indenter and rock are given in Table 2.2. The applied pressure on the indenter was 10 kgf/cm² (1 MPa) and the top and side of the model were assumed to be free to move while the central vertical line and bottom of the model can move vertically and radially, respectively.

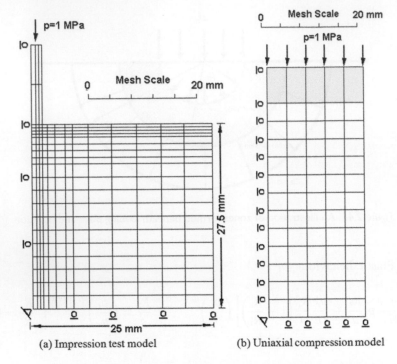

(a) Impression test model   (b) Uniaxial compression model

Figure 2.47 Finite element models for impression and uniaxial compression tests.

Table 2.2 Material properties used in finite element
analyses.

Material	Elastic modulus (GPa)	Poisson's ratio
Indenter	200	0.30
Rock	10	0.25

Similarly, the uniaxial compression experiment was analysed using the axisymmetric finite element method. The material properties used in the finite element analyses are the same as those used in the simulation of the impression model. The specimen is 12 cm high and 5 cm in diameter (Figure 2.47(b)). The platen was assumed to be 1 cm thick. One fourth of the sample is modelled using the symmetric characteristics of the problem.

Figure 2.48 shows the computed deformed configuration, minimum principal stress contours (tension is assumed to be positive) and maximum shear stress distribution. The minimum principal stress and maximum shear stress contours resemble to pressure bulbs as expected. However the distributions are not uniform just below the indenter. Nevertheless, the pressure bulbs become spherical beyond a distance

(a) Deformed configuration

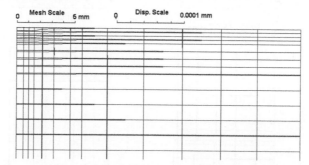

(b) Minimum principal stress contours

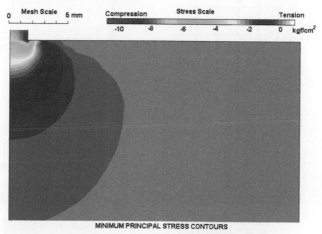

(c) Maximum shear stress

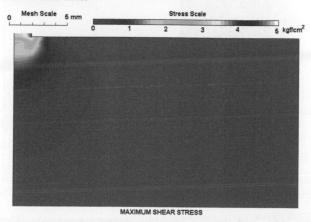

Figure 2.48 Computed results from the axisymmetric finite element analysis for impression test.

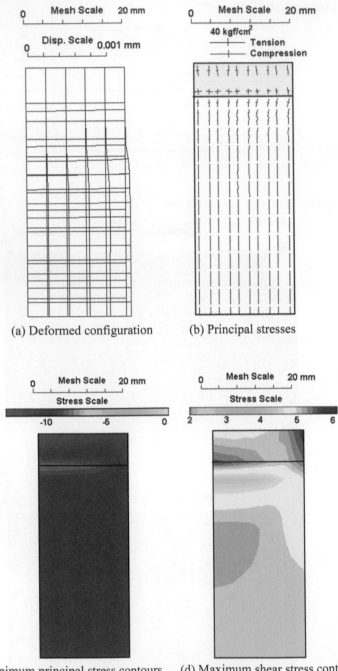

(a) Deformed configuration   (b) Principal stresses

(c) Minimum principal stress contours   (d) Maximum shear stress contours

Figure 2.49 Computed FEM results for uniaxial compression experiments.

Table 2.3 Computed average impression displacement.

Model	Computed displacement (mm)
FEM	1.815×10^{-4}
Eq. (2.103)	2.209×10^{-4}
Eq. (2.106) with Eq. (2.110)	1.875×10^{-4}

equivalent to the radius of the indenter. Table 2.3 compares the computed average displacement responses of the indenter from the finite element analysis and Eqs. (2.102) and (2.106) with (2.110). As noted from Table 2.3, the finite element analysis yields smaller displacement compared to the theoretical derivations. The reason for the discrepancy is due to the differences in boundary conditions. While the domain is finite in the FEM analysis, it is a half-space in the derivations of Eqs. (2.103) and (2.106). Nevertheless, it is interesting to note that Eq. (2.106) yields reasonably close results to those from the FEM, as the ratio of diameter and length of the domain is greater than 16 times the indenter radius.

Brazilian test is also carried out to infer the tensile strength of rocks under compression loading. It is theoretically derived that the tensile strength of rocks can be obtained. Tensile stress induced in a solid cylinder of rock is given by

$$\sigma_t = \frac{2F}{\pi D t} \tag{2.111}$$

where F, D and t are applied load, diameter and thickness of rock sample, respectively. The nominal strain of the Brazilian tensile test sample may be given as (see Hondros, 1959; Jaeger & Cook, 1979 for details)

$$\varepsilon_t = 2\left[1 - \frac{\pi}{4}(1 - v)\right]\frac{\sigma_t}{E} \quad \text{with } \varepsilon_t = \frac{\delta}{D} \tag{2.112}$$

For most rocks, the formula given above may be simplified to the following form

$$\varepsilon_t = 0.82\frac{\sigma_t}{E} \tag{2.113}$$

A plane stress finite element analysis was carried out for the Brazilian test. The properties of rock were the same as those given in Table 2.2. The properties of the platen were assumed to be those of the aluminium with an elastic modulus of 70 GPa. Uniform compressive pressure with an intensity of 20 kgf/cm^2 was applied on the platens and boundary conditions are shown in Figure 2.50.

The maximum tensile stress occurs in the vicinity of the centre of the sample and its value is 1.08 kgf/cm^2. This is slightly greater than the theoretical estimation of 0.8 kgf/cm^2. This is probably due to the slight difference in the application of load boundary conditions. The radial displacement of the sample just below the platen is

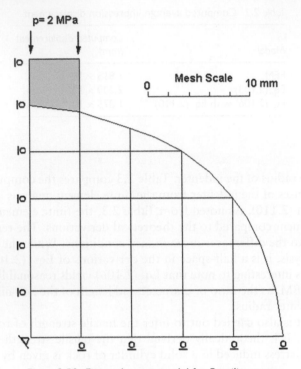

Figure 2.50 Finite element model for Brazilian tests.

about 0.001 mm, which is almost equal to that estimated from Eq. 2.113. Therefore, it is possible to determine the elastic modulus besides the tensile strength of rocks. Furthermore, the strain response in creep experiments should be similar to those of the uniaxial compression creep experiments provided that deformability characteristics remain the same under both tension and compression.

2.7 CREEP EXPERIMENTS ON OYA TUFF

2.7.1 Geology and stability problems of underground quarries in Oya region

Furthermore, Oya tuff or Oya stone, which is one of the most popular building stone materials in Japan, has been quarried in the Oya region, Utsunomiya, Japan. Over 200 underground quarries have been exploited for more than 120 years and some of those are below residential zone. The Oya tuff that is quarried at a hilly region nearby Oya town, Utsunomiya City, Tochighi Prefecture, Japan with an elevation of 200 m (Figure 2.52). It belongs to Ashio belt and it is a Tertiary formation with the basement rocks such as chert, sandstone of the Paleozoic era, and the Mesozoic era and covered with the diluvium of the Quaternary period, the conglomerate layer and Kanto loam

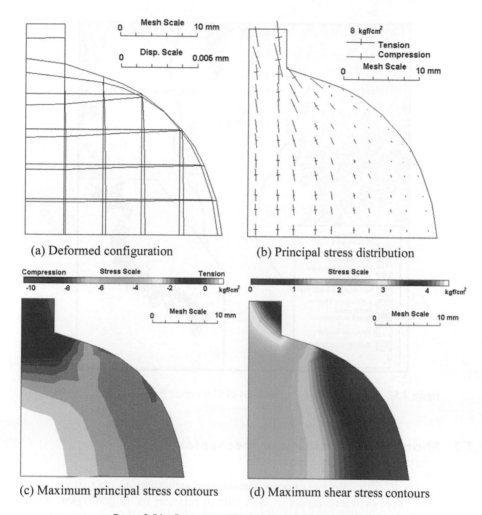

(a) Deformed configuration   (b) Principal stress distribution

(c) Maximum principal stress contours   (d) Maximum shear stress contours

Figure 2.51 Computed FEM results for Brazilian tests.

layer. The Oya tuff was formed under marine environment about 20 million years ago and has a porous structure and bluish-green pumice in splashed patterns together with chunks of the clay mineral. Its clay mineral mainly consists of montmorillonite and zeolite. Oya tuff is a soft rock and it can be easily excavated. However, it is easily weathered and degraded. Especially the chunks of clay mineral in Oya tuff are easily washed away.

Underground quarries are exploited using the room and pillar method. There were 15 large-scale collapses in Oya region since 1946 (Figure 2.53). Oya tuff contains some swelling clay minerals (Table 2.4) and it is susceptible to swelling and shrinkage (Figure 5.2). Free swelling strain of Oya tuff is generally less than 0.6%.

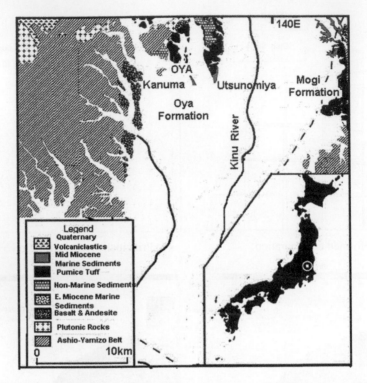

Figure 2.52 Location and geological maps of Oya region (from Aydan *et al.*, 2006).

2.7.2 Short term physical and mechanical properties of Oya tuff

Depending upon sedimentation process, the internal structure of Oya tuff varies. There are basically three different layers of Oya tuff in relation to its geological past. The grain size of Oya tuff changes and the highest strength obtained for fine-grain Oya tuff while lowest strength is obtained for coarse grain type of Oya tuff. Therefore, the scattering of material properties is large as noted from Table 2.5. Furthermore, its strength and deformability properties change in relation to water content (Figure 2.55). The reduction of properties (ϕ) are fitted to the following function proposed by Aydan & Ulusay (2002):

$$\phi = \phi_0 - (\phi_{100} - \phi_0)\frac{S}{S + \beta(100 - S)} \qquad (2.114)$$

where S is saturation ranging between 0 (dry) and 100 (fully saturated). ϕ_0 and ϕ_{100} are the normalized values of properties for the saturation values of 0 and 100, respectively. β is an empirical constant. The value of β is 0.2 for properties shown in Figure 2.55.

A series of short term experiments on Oya tuff samples were carried out on samples of Oya tuff, which are also used in creep experiments reported in this section.

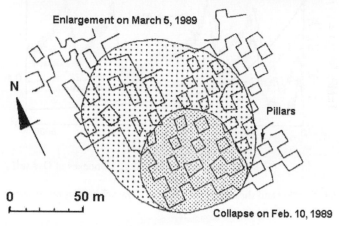

Enlargement on March 5, 1989

N

Pillars

0   50 m

Collapse on Feb. 10, 1989

Figure 2.53 A view and pillar layout at a large-scale collapse of underground quarry in Oya region.

Table 2.4 XRD results from the samples of Oya tuff (from Aydan *et al.*, 2006).

Specimen Number	Whole Rock					Clay fraction		
	Quartz	Opaque-CT	Feldspar	Dolomite	Clay	Smectite	Kaolin	Illite
Oya	7.3	29.5	4.9	–	33.3			

The uniaxial compressive strength and Brazilian tensile strength of Oya tuff samples are summarized in Table 2.6. Figure 2.56 and Figure 2.57 shows the multi-parameter responses of Oya samples during a Brazilian test and a compression test. Figure 2.58 show the responses of during penetration experiments with an indenter having a diameter of 3 mm under dry and saturated conditions together with estimated responses and

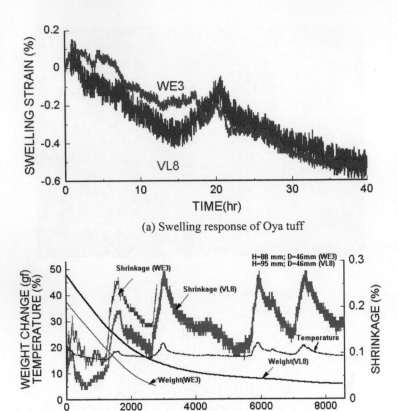

(a) Swelling response of Oya tuff

(b) Shrinkage and weight change responses of Oya tuff

Figure 2.54 Swelling and shrinkage responses of Oya tuff in relation to water content.

Table 2.5 Mechanical characteristics of dry tuffs of Oya (from Aydan *et al.*, 2006).

Parameter	Oya
Dry unit weight (kN/m³)	13.5–15.7
Saturated unit weight (kN/m³)	17.0–17.6
Effective porosity (%)	23.4–32.0
Uniaxial compressive strength (MPa)	5.72–24.8
Tensile strength (MPa)	0.82–1.52
Elastic modulus (GPa)	0.90–3.00
Poisson's ratio	0.25–0.30
Friction angle (°)	27–39
P-wave velocity (km/s)	1.6–1.8
S-wave velocity (km/s)	
Thermal conductivity coefficient (kcal/h m °C)	0.7–1.3
Thermal expansion coefficient (×10⁻⁶ 1/°C)	8.0
Hydraulic conductivity (×10⁻⁶ cm/s)	

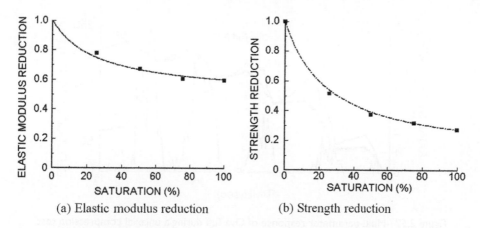

(a) Elastic modulus reduction   (b) Strength reduction

Figure 2.55 Elastic modulus and uniaxial compressive strength reduction in relation to water content variation (saturation).

Table 2.6 Short term properties of Oya tuff used in creep experiments.

Condition	Compressive strength (MPa)	Tensile Strength (MPa)	Elastic modulus (GPa)
Dry	5.89–9.51	0.71–0.94	0.7–1.63
Saturated	2.01–3.98	0.25–0.38	0.34–0.61

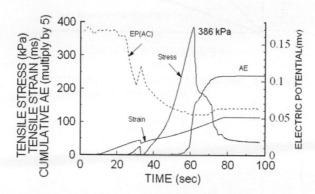

Figure 2.56 Multi-parameter response of Oya Tuff during a Brazilian test.

properties from the theory of Aydan *et al.* (2008) for penetration experiments. It is also possible to determine the triaxial strength of rocks using the results of the Brazilian tensile strength experiment and uniaxial compression experiments. Besides the Mohr-Coulomb yield criterion, Aydan modified his criterion by relating its parameters to the parameters of Mohr-Coulomb criterion for isothermal conditions as follows:

$$\tau = c_\infty - (c_\infty - c_o)e^{-\sigma_n/b} \tag{2.115}$$

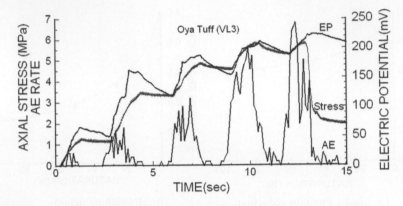

Figure 2.57 Multi-parameter response of Oya Tuff during a uniaxial compression test.

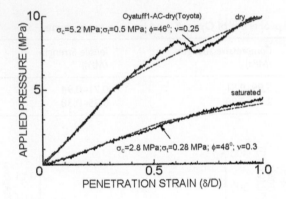

Figure 2.58 Response of Oya Tuff during a penetration test.

where

$$b = \frac{\sigma_t}{\ln\left(\frac{c_\infty}{c_\infty - c_o}\right)}$$

Table 2.7 gives the parameters of Mohr-Coulomb and Aydan's failure criteria. Figure 2.59 compares the yield criterion for dry and saturated conditions.

2.7.3 Brazilian tensile creep experiments

A specially designed cantilever type device used for Brazilian creep tensile creep experiments. This system induces loads on samples six times the load imposed at the cantilever end. Diameter of samples is 46 mm and their thickness ranges between 14 and 25 mm. The monitoring system involves the measurement of displacement, acoustic emission

Table 2.7 Short term triaxial strength parameters of Oya tuff used in creep experiments.

Parameter	Dry		Saturated	
	Lower	Upper	Lower	Upper
c_0 (MPa)	0.35	0.70	0.35	0.7
c_∞ (MPa)	12.0	16.0	4.0	8.0
ϕ (°)	51	55	51	55
b (MPa)	8.8	10.0	2.7	4.73

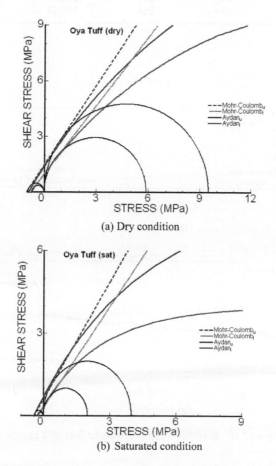

(a) Dry condition

(b) Saturated condition

Figure 2.59 Illustration of shear strength of Oya-tuff under dry and saturated conditions.

counts and electrical potential. The monitoring system operates entirely on battery-operated monitoring device. Therefore, the experimental set-up can be easily used under in-situ conditions without any modifications provided that sufficient protection against vandalism by humans and/or nature is implemented. Table 2.8 gives the

Table 2.8 Conditions and measured parameters of Brazilian creep experiments.

Sample No.	1st loading (kPa)	2nd loading (kPa)	3rd loading (kPa)	Loading period (min)	Failure time (min) under saturation
SN1-W3	125	168	335	4306	1278
WE4N-3	156	302	363	5448	6
VS5	82	163	245	11352	8472
SN5E-E	415	–	–	1465	3
VL15-B1	135	224	269	4237	1579
VL15-B2	138	231	323	4023	35
WE1S-D	150	250	–	10066	8482 (no failure)
WEZ-3	227	247	269	22594	19594 (no failure)
EW4S-5	116	194	256	10500	8700 (no failure)

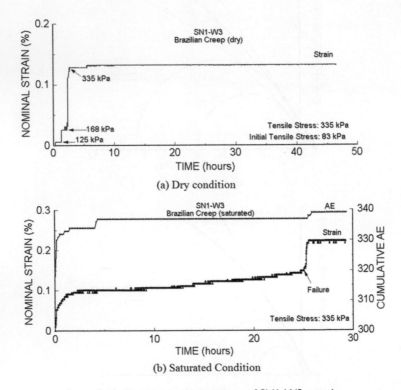

(a) Dry condition

(b) Saturated Condition

Figure 2.60 Brazilian creep response of SN1-W3 sample.

conditions and measured parameters of experiments. All samples were subjected creep experiments up to a chosen period of time under dry conditions and then it was fully saturated. Some creep experiments under dry condition were performed at Toyota National College of Technology. Figures 2.60 to 2.68 shows the measured responses of samples of Brazilian creep experiments.

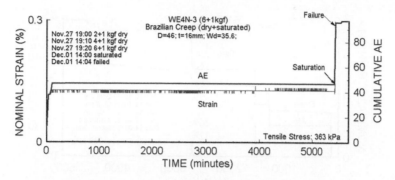

Figure 2.61   Brazilian creep response of WE4N-3 sample.

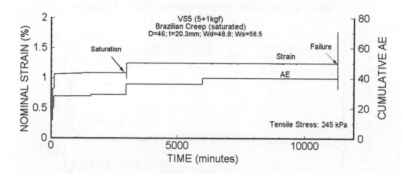

Figure 2.62   Brazilian creep response of VS5 sample.

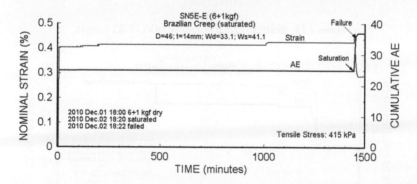

Figure 2.63   Brazilian creep response of SN5E-E sample.

2.7.4   Impressions creep experiments

The device shown in Figure 2.14 has been used for experiments reported in this section. The diameter of the indenter is 3 mm and the device is of cantilever type and it is capable of inducing 10 times the applied load at the end of the arm. The device equipped with a displacement transducer and an acoustic emission sensor.

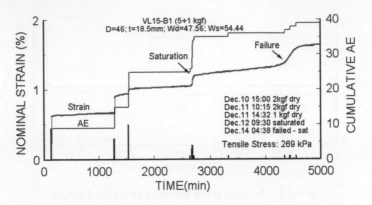

Figure 2.64 Brazilian creep response of VL15-B1 sample.

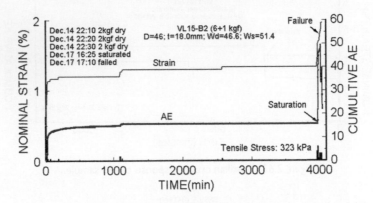

Figure 2.65 Brazilian creep response of VL15-B2 sample.

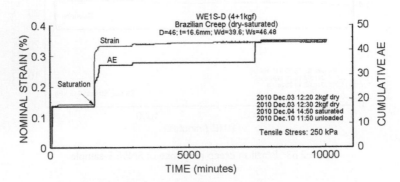

Figure 2.66 Brazilian creep response of WE1S-D sample.

The monitoring is entirely based on battery-operated loggers as used in Brazilian creep experimental device. Figures 2.69–2.73 shows the responses measured in impression experiments. Table 2.9 gives the details of experimental conditions.

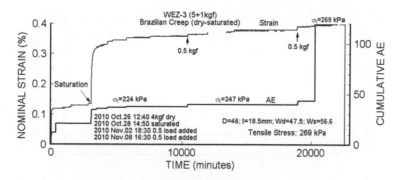

Figure 2.67   Brazilian creep response of WEZ-3 sample.

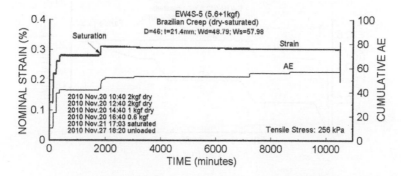

Figure 2.68   Brazilian creep response of EW4S-5 sample.

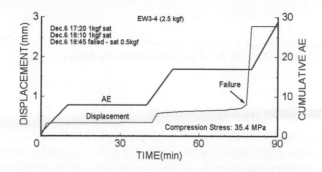

Figure 2.69   Impression creep response of EW4S-5 sample.

2.7.5   Uniaxial creep experiments

The device shown in Figure 2.6 has been used for experiments reported in this section. The device is of cantilever type and it is capable of inducing loads up to 50 kN. The device equipped with a displacement transducer. Load cell and an acoustic emission

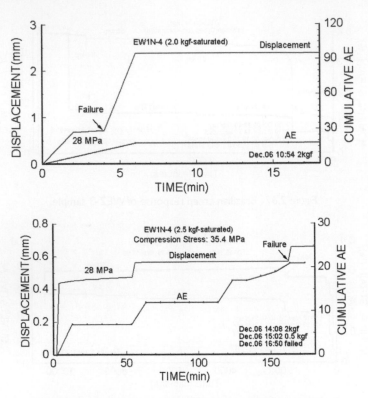

Figure 2.70 Impression creep responses of EWIN-4 sample.

sensor. Figure 2.74 shows the responses measured in uniaxial compression experiments so far. Table 2.10 gives the details of experimental conditions.

2.7.6 Comparisons of experiments

Figure 2.75 compares the failure time of samples tested in Brazilian, impression and uniaxial compression creep experiments under dry and saturated conditions. Although it is difficult to compare the impression creep experiments, the stress ratio is obtained by dividing the applied stress condition by 36 MPa. This is a first trial and a short term penetration experiments under dry and saturated conditions are felt to be necessary for better comparative results. From experimental results, it is very interesting to note that if the stress ratio remains same, the failure time of dry and saturated samples are very close to each other. Furthermore, the failure times of samples tested under uniaxial compression and Brazilian creep experiments are also similar. These two important conclusions have strong implications in practice. This comparison is still preliminary and further comparisons of experimental results are needed.

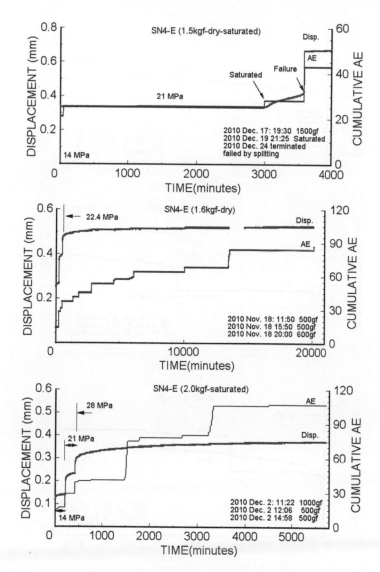

Figure 2.71 Impression creep responses of SN4-E sample.

2.8   APPLICATIONS OF THE LONG TERM RESPONSE AND STABILITY OF ROCK ENGINEERING STRUCTURES

2.8.1   Abandoned room-pillar mines

One of the main parameters influencing the long term stability of abandoned mines is the creep strength characteristic of rocks. A series of creep tests on sandstone samples

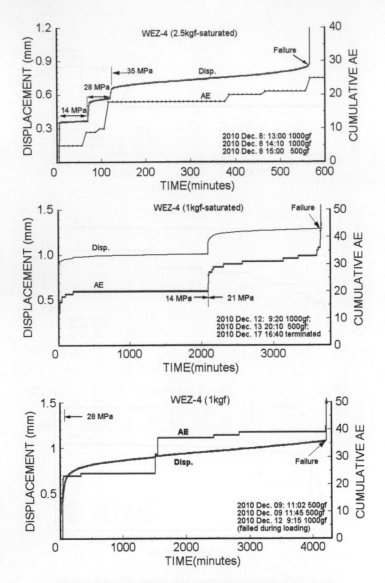

Figure 2.72 Impression creep responses of WEZ-4 sample.

obtained from abandoned lignite mines of Mitake town in Gifu Prefecture of Japan were carried out. The specific of the function (2.63) proposed by Aydan & Nawrocki (1998) for the long term strength of rocks normalized by their short term strength is adopted for the long term stability analyses of abandoned room and pillar mines together with the use of RMR rock classification system (Aydan *et al.*, 2005).

$$\frac{\sigma_{cl}}{\sigma_{cs}} = 1 - 0.0397 \ln\left(\frac{t}{t_s}\right) \tag{2.116}$$

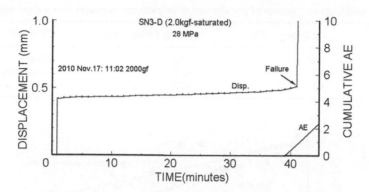

Figure 2.73 Impression creep response of SN3-D sample.

Table 2.9 Conditions and measured parameters of Impression creep experiments.

Sample No.	1st loading (MPa)	2nd loading (MPa)	3rd loading (MPa)	Loading period (min)	Failure time (min) under saturation
EW3-4	14.14	28.28	35.35	80	80
EW1N-4	28.28	35.35	–	164	164
SN4-E-1	21.21	–	–	3576	585
SN4-E-2	22.4	–	–	20084	–
SN3-D	28.28	–	–	40	40
WEZ-4-1	14.14	28.28	–	3664	440
WEZ-4-2	14.14	21.21	323	4023	1573
WEZ-4-3	14.14	–	–	4200	–

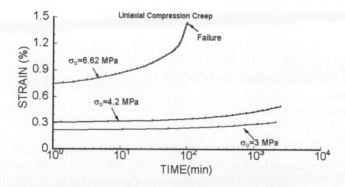

Figure 2.74 Uniaxial creep responses of samples.

Table 2.10 Conditions and measured parameters of uniaxial compression creep experiments.

Sample No.	1st loading (MPa)	Loading period (min)	Comment
VU3	6.62	100	failed
EW4S	4.2	2458	Terminated – no failure
WEIS	3.0	2130	Terminated – no failure

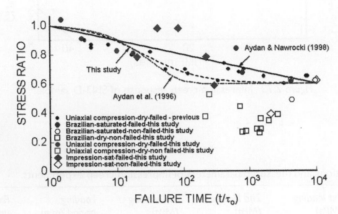

Figure 2.75 Comparison of failure time of Brazilian creep samples with that of uniaxial compression creep experiments.

For the simplicity, we will use the tributary area approach for pillar stability and arching approach for the stability of roof layers in developing our theoretical equations. The tributary area is simple yet effective method to assess the overall stability of the pillars. Let us assume that the geometry of a representative pillar and its overburden load can be modeled as shown in Figure 2.76(a). From the assumed geometry, it is easy to show that the average pillar stress is

$$\sigma_p = \rho g H \frac{A_t}{A_p} \quad \text{or} \quad \sigma_p = \sigma_V \frac{A_t}{A_p} \tag{2.117}$$

where
ρ: is unit weight of rock
g: is gravitational acceleration
H: Overburden
A_t: Area supported by pillar
A_p: Area of pillar
$\sigma_V = \rho g H$

The long term strength of rock depends upon the level of the sustained loading with respect to its strength in short term tests. Figure 2.76(b) shows the computed results for the failure time of pillars with the consideration of long term strength of pillars. From the results it is inferred that the failure time would be shorter if the

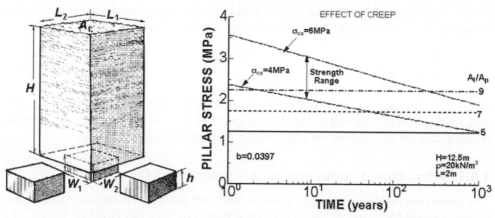

(a) Tributary area model   (b) Estimated collapse time of abandoned lignite mines

Figure 2.76 Mechanical model and failure time of pillars of abandoned lignite mine.

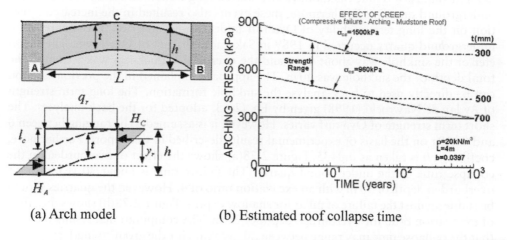

(a) Arch model   (b) Estimated roof collapse time

Figure 2.77 Mechanical model and failure time of roof of abandoned lignite mine.

tributary area (excavation ratio) increases. This type of situations may be observed when the abandoned mines are fully submerged. However, it should be noted that the long term strength properties for saturated conditions should be used under such situations. The range probably corresponds to the actual situation of the abandoned mine that the author and his group have been investigating (Aydan *et al.*, 2005). For the lower bound strength range, the stability problems of pillars would occur after 57 years for a typical excavation ratio of 7. The recent collapses of abandoned lignite mines in the town of Mitake confirm this conclusion.

The stability of roof layers can be checked against various forms of failure. Although roof layers may crack due to bending stresses, their ultimate failure would be governed by arching phenomenon within the roof layers. If the fiber stress exceeds the compressive strength at the crest of arch within roof layer, the roof layer will collapse. Under such a condition, the following relation can be derived (i.e. Aydan, 1989):

$$\sigma_c = \frac{2}{3}\rho g h \left(\frac{L}{h}\right)^2 \tag{2.118}$$

where L and h are the span and thickness of roof layer, respectively. The effect of degradation can be imposed on the thickness of roof layer while the effect of long term strength can be imposed on the compressive strength of roof layers.

2.8.2 Abandoned room and pillar quarries of Oya tuff

The quarrying history of Oya tuff is more than 100 years and it became famous as a building stone following its use in the construction of Imperial Hotel in Tokyo. The mechanical excavation using chains started in 1952 and the size of quarries become larger as a result. The first caving of the quarries occurred in 1946 and there were very large scale caving (sinkholes from 1989 to 1991, which received wide coverage by the mass media. As a result, Utsunomiya city started long term monitoring of abandoned underground quarries. Furthermore, these events also resulted in the increased attention on the long term stability of Oya tuff underground quarries. The failure of an underground quarry occurred in 1989 in Sakamoto district of Oya town. The diameter of the sinkhole was about 65 m and the overburden thickness was 30 m and the total depth of the sinkhole was 40 m. The approach explained in the previous section can be directly used to back-analyse the sinkhole formation. The long term strength of Aydan & Nawrocki (1998) given by Eq. (2.63) adopted for the back-analyses. The short term strength of Oya tuff varies. However, it is assumed to be ranging between 6 and 8 MPa on the basis of experimental results described in this report. Furthermore, coefficient b is taken as 0.0435. Figure 2.78(a) shows the effect of overburden on the collapse time of the underground quarry. The failure time is about 30 years for an overburden depth of 30 m with an excavation ratio of 4. However, the quarries should be stable against the failure of pillar for shallow depths. Figure 2.78(b) shows the effect of excavation ratio for overburden depth of 30 m. The computational results indicate that the collapse time may range between 20–30 years for the given strength range and chosen excavation ratios.

2.8.3 Man-made natural underground openings in Cappadocia region

(a) Stability of openings to next to cliffs and in fairy chimneys

Based on the observations of the Aydan *et al.* (1997, 2008) at Göreme, Zelve, Ürgüp, and Uçhisar and Ortahisar castles, most of the instabilities occur as a result of collapse of toes of openings next to cliffs. In such failures, erosion of the toe by natural agents and decrease in long term strength of the rock are the main causative factors. In addition, instability problems were also observed in openings in fairy chimneys. In the model representing an opening next to the cliff (Fig. 2.79a), it is assumed that the pillar or wall at the valley side carries the half of the opening. Due to its conical

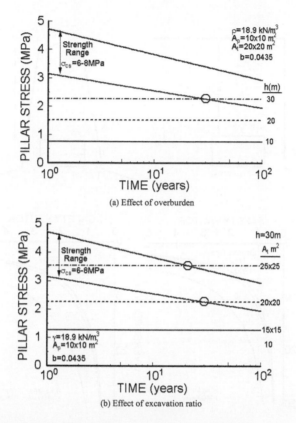

Figure 2.78 The effect of overburden and excavation ratio on the failure time of collapsed region of Sakamoto underground quarry in Oya town.

shape, a fairy chimney is considered as an axially symmetrical rock structure including a circular opening at its center (Fig. 2.79a). Although the real stress distributions in these openings are slightly different than those in these models, it is considered that the approaches used will be helpful to assess the conditions of the instabilities investigated. The time dependent safety factor (SF) of the wall next to the cliff and in fairy chimney is written as follow.

$$SF = \frac{\sigma_{cr}(t)}{\gamma H} \frac{A_w}{A_t} \qquad (2.119)$$

where γ, σ_{cr}, H, A_t and A_w are unit weight, creep strength, overburden height, total area supported and wall area. The area ratios for continuous wall next to cliff and in cylindrical fairy chimney shown in Figure 6.4a specifically take the following forms, respectively.

$$\frac{A_t}{A_w} = 1 + \frac{w}{2B} \qquad (2.120a)$$

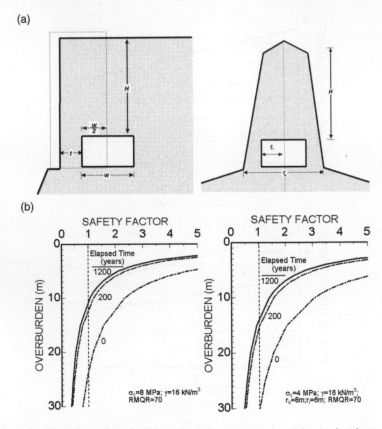

Figure 2.79 (a) Mechanical models and (b) safety factor variation with overburden and time for openings next to cliffs and in fairy chimneys, respectively.

$$\frac{A_t}{A_w} = \frac{1}{1 - \left(\frac{r_i}{r_0}\right)^2} \tag{2.120b}$$

The creep strength (σ_{cr}) in terms of short term strength (σ_{cs}) is represented in the following form for tuffs of Cappadocia region.

$$\sigma_{cr} = \sigma_{cs}\left(1 - 0.05\ln\left(\frac{t}{\tau}\right)\right) \tag{2.121}$$

where: t and τ are time and short term test duration, respectively. The uniaxial compressive strength of the rock mass is estimated from the equation of Aydan *et al.* (2014):

$$\frac{\sigma_{cm}}{\sigma_{ci}} = \frac{RMQR}{RMQR + \beta(100 - RMQR)} \tag{2.122}$$

The value of β in equation above can be taken as 6 on the basis of experimental data from construction sites in Japan (Aydan & Kawamoto, 2000; Aydan *et al.*, 2014). Equation 2.122 can be adopted for the creep strength versus failure time function and the safety of openings can be evaluated. Figure 2.79b shows the safety factor of the openings next to cliffs and in fairy chimneys as a function of time and overburden for strength properties of tuffs. It is clear from Figure 2.79b that openings next to cliffs are more likely to fail in long term as compared with those of fairy chimneys.

(b) Long term stability of Derinkuyu underground city

In the elasto-visco-plastic analyses, the total strain of surrounding rock mass consists of instantaneous strain due to excavation and creep strain.

$$\varepsilon_t = \varepsilon_e + \varepsilon_c \tag{2.123}$$

The instantaneous strain field is obtained from the elastic or elasto-plastic behaviour of rock mass. The creep strain is obtained from visco-elastic or visco-elasto-plastic behaviour following the excavation. This mechanical model can be expressed for linear behaviour in the following form

$$\sigma = D_e \varepsilon \quad \text{at the time of excavation} \tag{2.124}$$

$$\sigma = D\varepsilon + C\dot{\varepsilon} \quad \text{after the excavation} \tag{2.125}$$

This model is fundamentally very similar to the Generalized Kelvin Model or Zener Model. In the finite element formulation of rock excavations, multi-dimensional forms of Generalized Kelvin Model or Zener Model are used. If such a concept is formulated using the finite element method, one can easily get the following form:

$$C\dot{U} + KU = F \tag{2.126}$$

where

$$C = \int_\Omega B^T CB d\Omega; \quad K = \int_\Omega B^T DB d\Omega; \quad F = \int_\Omega N^T b d\Omega + \int_\Gamma N^T b d\Gamma$$

At time zero (excavation), the initial strain field is obtained using the multi-dimensional elasticity tensors of either Zener or Generalized Kelvin Model and Eq. (2.38) is integrated over the time domain using one of time-integrations schemes. When the behaviour deviates from linear visco-elasticity behaviour, the visco-plastic schemes can be implemented as described in a textbook by Oven & Hinton (1980).

The models described in previous section for one-dimensional situation can also be extended to one-dimensional axi-symmetric situations. Figure 2.80 compares the finite-element analysis based on the generalized Kelvin model with analytical results for one-dimensional axi-symmetric situation for the shaft of Derinkuyu underground city.

A series of visco-elasto-plastic finite element analyses were carried out to assess the short and long term stability of a vertical shaft and a hall at the 7th floor of Derinkuyu Underground City (Aydan *et al.*, 2008; Aydan & Ulusay, 2013). The vertical shaft is still in operation and is used as one of the ventilation shafts connecting all floors of

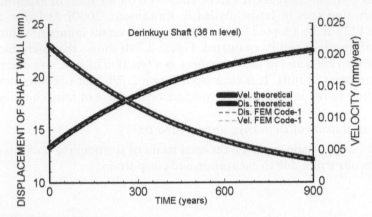

Figure 2.80 Comparison of analytical solutions with axisymmetric finite element solution.

Table 2.11 Physical and mechanical properties used in finite element analyses.

ρ kN/m³	λ_0 MPa	μ_0 MPa	λ MPa	μ GPa	λ^* GPa year	μ^* GPa year	ϕ (°)
18	121	121	69	69	0.242	0.138	22

Derinkuyu Underground City to the ground surface. The cross-section of the shaft is circular. Therefore, the problem was treated as an axi-symmetric problem. The shaft was taken as 37.5 m deep and the hall was at the depth of 40 m. The vertical stress was assumed to be equivalent to the overburden and the lateral stress coefficient was taken 0.5 on the basis of initial in-situ stress predictions (Aydan *et al.*, 1999; Watanabe *et al.*, 1999; Aydan & Ulusay, 2013). The Mohr-Coulomb yield function was replaced with Drucker-Prager yield criterion by using the theoretical relations between two yield criteria while taking into account the strain rate dependency. Relevant mechanical properties used in the analyses are given in Table 2.11, which were determined from short and long term laboratory tests on rock samples and in-situ characterization of rock mass.

The analysis on the shaft showed that there should be no yielding occurring around the shaft soon after the excavation. The computations were carried out up to 1500 years, which corresponds to the present time if this underground city was assumed to be excavated 1500 years ago. Figure 2.81 shows the maximum shear stress distribution contours at the time of 1500 years following the excavation and also the displacement and velocity responses of the shaft at the level of 36 m from the ground surface. Computations also indicated that yielding of rock mass should not also occur at the present time. Since the behaviour of surrounding rock mass is visco-elastic, the deformation of the shaft should had been nearly converged to its final value and the stress state in the surrounding rock mass should also be the same as that at the time of excavation.

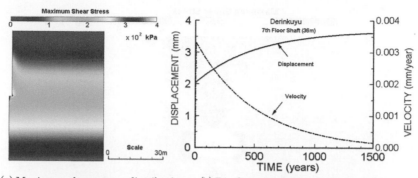

(a) Maximum shear stress distributions   (b) Displacement and velocity responses

Figure 2.81   Computed stress distributions and displacement and velocity responses of shaft wall at 36 m level using elasto-visco-plastic FEM.

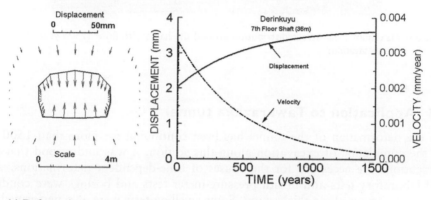

(a) Deformation vectors at 1500 years   (b) Displacement and velocity response

Figure 2.82   Deformation and velocity responses of the hall at 7th floor of Derinkuyu underground city.

The second example is concerned with the hall at the 7th floor of Derinkuyu Underground City. The hall has an arch shaped roof and its shape is close to horseshoe. The width is about 4.5 m and 20 m long. Therefore, the problem was treated as a two dimensional plain-strain problem. Although the hall has 3 pillars along the centre line, they were neglected in computations as they had some thoroughgoing discontinuities. The analysis showed that no yielding occurs soon after the excavation. The computations were carried out up to 1500 years, which corresponds to the present time. Figure 2.82 shows the deformed configurations of the domain analysed 1500 years after the excavation and the displacement and velocity response of the hall. Computations also indicated that yielding of rock mass should not occur until the present time. Since the behaviour of surrounding rock mass is visco-elastic, the deformation of the shaft should had been nearly converged to its final value and the stress state in the surrounding rock mass should also be the same as that at the time of excavation. Figure 2.83 shows the maximum shear stress state around the hall.

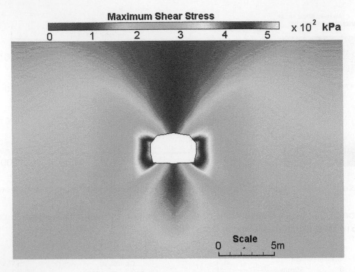

Figure 2.83 Maximum shear stress distribution around the hall at 7th floor (1500 year elapsed after the excavation).

2.8.4 Application to Tawarazaka tunnel

Since the deformation of the tunnel has been continuing for more than 1500 days after the completion of excavation along this section, it was understood that some investigations are necessary for the causes of time-dependent deformations. Additional laboratory tests and in-situ pressure-meter tests and borings were conducted at certain locations along this section. Some swelling tests were also performed. The swelling tests indicated that rocks in this section had no swelling potential. This was unexpected.

(a) Simplified Time-dependent axi-symmetric analyses

Aydan *et al.* (1992, 1993, 1996) proposed some analytical solutions for predicting the deformation response of tunnels in squeezing rocks. Using the concept proposed by Ladanyi (1974), Aydan *et al.* (1996) considered the variation of long term properties of rocks with time and they called as the degradation of material properties. This concept was adopted in the original model of Aydan *et al.* (1992, 1993) and it was applied at a section of tunnel where the overburden was 280 m. The three different long term strength values were chosen and deformation responses of the tunnel were computed for each long term strength values. Figure 2.84 shows the computed deformation responses with the measurements at three sections. As noted from the figure, the deformation response of the tunnel implies remarkable visco-elastic behaviour. It is also interesting to note that when the long term strength if 0.5 times the short term strength, there is an abrupt variation of the deformation response as soon as rock yields. This response is remarkably similar to those measured.

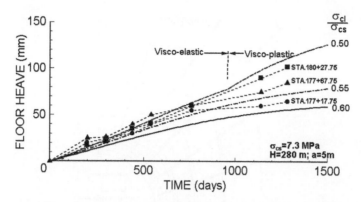

Figure 2.84 Comparison of computed responses with measurements.

Table 2.12 Physical and mechanical properties used in finite element analyses.

ρ kN/m³	λ_0 MPa	μ_0 MPa	λ^* MPa day	μ^* MPa day	α_{st}	α_{lt}	K_{st} MPa	K_{lt} MPa
22	275	254	3.96	3.65	1.51	0.9	0.19	0.16

(b) Elasto-visco-plastic finite element analyses

The same section of the Tawarazaka tunnel was analysed using the elasto-visco-plastic finite element method (Aydan *et al.*, 1995). The finite element analyses were also concerned with the effect of support system. First, unsupported case is considered. The overburden was assumed to be 280 m at the respective location. The physical and mechanical characteristics of rock are given in Table 2.12. Figure 2.85 shows the deformed configuration of surrounding rock around the tunnel at 2500 days after excavation.

Figure 2.86 shows the finite element mesh for supported case in which bold lines correspond to bolts, shotcrete and steel ribs. This support system was initially adopted for supporting the tunnel through this section. Rockbolts and steel ribs and shotcrete were represented by rockbolt element (Aydan, 1989), shotcrete element (Aydan *et al.*, 1990), respectively. Figure 2.87 shows the deformed configuration of surrounding rock around the tunnel at 2500 days after excavation.

Figure 2.88 compares the time history of displacements in the centre of the floor after excavation (note that displacements due to excavation were subtracted from the figures) for supported and unsupported cases together with measurements at the centre of the floor. While the time dependent deformation of the crown decreased from 48.9 mm to 33.7 mm at 2500 days due to the effect of support pattern, no remarkable difference was observed for the floor as shown in Figure 2.88. This was thought to be due to the pattern of the support system since it was a non-closed support ring.

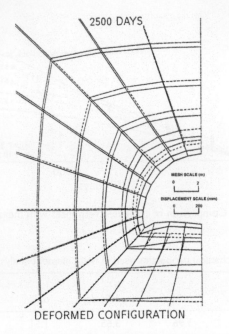

Figure 2.85 Computed deformed configuration.

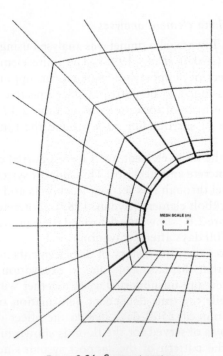

Figure 2.86 Support system.

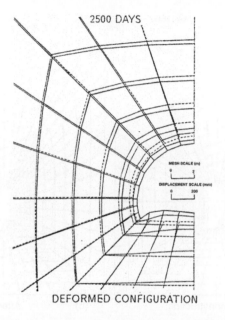

Figure 2.87 Computed deformed configuration (with support) responses with measurements.

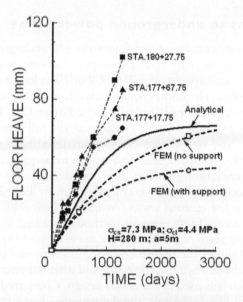

Figure 2.88 Comparison of computed.

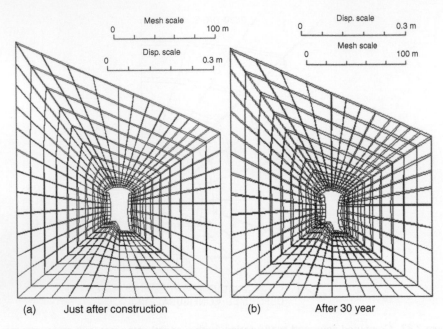

Figure 2.89 Comparison of deformed configuration just after construction and 30 years.

2.8.5 Applications to underground power house

A pumped-storage scheme consists of two reservoirs and underground powerhouse and it was constructed about 30 years ago. The upper reservoir consists of a 125 m high rockfill-dam with a storage capacity of 5,780,000 m^3. The lower reservoir has a gravity dam with a height of 44 m. The underground powerhouse is 55 m long, 22 m wide and 39 m high and it has two turbines (Chubu Electric Power Co., 1979). The maximum water level variation may reach 45 m in 12 hours at the full capacity. As it is expected that the deformability and strength of every geomaterial have some time-dependent characteristics, the time-dependent behaviour of the underground power house would naturally occur following its construction. The time-dependency of cohesion (c) of rock mass was assumed to obey the functional form given by Eq. (2.101) by considering experimental results on the igneous rocks (Aydan & Nawrocki, 1998).

The elasto-visco-plastic finite element method described in previous section was used to analyse the long term response of rock mass around the underground cavern. Figures 2.89 and 2.90 shows the deformed configurations and plastic zone formation around the cavern at the time of construction and after 30 years for the lateral stress coefficient of 1.0, which was used in the initial design. Compared with the deformation of the cavern shown in Figure 2.89(a), the deformation becomes larger. Furthermore, a plastic zone develops in the vicinity of the sidewall, which was not observed during the excavation step (Figure 2.90).

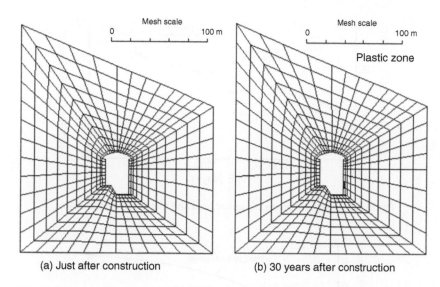

(a) Just after construction   (b) 30 years after construction

Figure 2.90   Comparison of plastic zone formation around the cavern just after and 30 years after the construction.

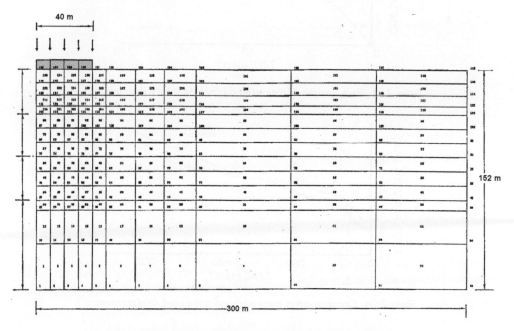

Figure 2.91   Finite element mesh used in the back-analyses.

2.8.6   Applications to foundations

The final application involves a back-analysis of the constitutive law parameters of the foundation rock of the pier 3P of the Akashi suspension bridge. The rock consists of

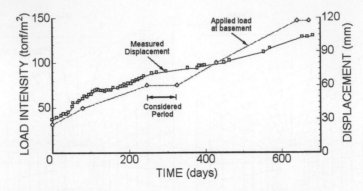

Figure 2.92 Time response of applied load and measured displacement.

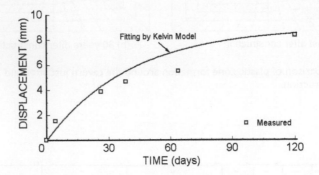

Figure 2.93 Back analysis of measured displacement by Kelvin model.

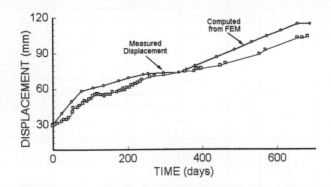

Figure 2.94 Comparison of measured and computed displacement.

Kobe tuff, which is a relatively soft-rock. The diameter of the foundation was 80 m and its height was 80. Following the lowering of the caisson foundation to the sea bottom, it was filled with concrete, which increased the load on the foundation. First the filling of the inner ring was completed and the outer ring following with some time lag. The deformation of the ground was measured during the filling stages. It was required to obtain the time-dependent characteristics of foundation formation by considering

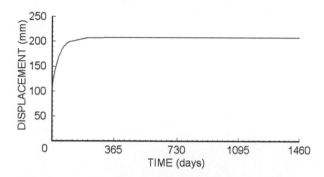

Figure 2.95 Estimated creep displacement of the pier.

the loading associated with the construction procedure. The constitutive law of the foundation rock was assumed to be of Kelvin type. The problem was considered to be an axi-symmetric problem and the finite element mesh used in the back analyses is shown in Figure 2.91. The elastic modulus and viscosity coefficient of Kelvin model of foundation rock were 833 MPa and 3.3 GPa day, respectively (Figures 2.92 and 2.93). Figure 6.94 compares the computed response with measured response for the loading condition shown in the same figure. Figure 6.95 shows the displacement of the pier for about 4 years. The expected creep displacement is about 108 mm.

Chapter 3

Water migration in soft rocks and its effects on the response of rock structures

3.1 INTRODUCTION

It is well known that clay-bearing and some evaporate rocks cause various engineering problems (Figure 3.1). Rocks such as mudstone, marl, siltstone, shale, tuff and weathered igneous rocks can be classified as clay bearing rocks. Most clay minerals exhibit volumetric variations when they absorb or desorb water. As a result, their physical and mechanical properties vary with the amount of water contained in such geomaterials.

(a) Underground room in Tomb of Ramses II

(b) Borehole in Tono URL

(c) Entrance gallery at Mitake Lignite Mine

(d) Unstable district In Babadağ

Figure 3.1 Examples of some engineering problems in rocks prone to water absorption/desorption.

Various structural and environmental problems in geo-engineering are caused by such volumetric variations. The mechanism associated with the mechanical weakening or softening of such geo-materials is thought to be due to the variation of the distance between sheets of clay minerals when clay fragments absorb or desorb water.

The slaking durability index test was proposed by Franklin & Chandra (1972) and adopted as suggested method by ISRM as an index test of rocks for characterization purposes. However, such an index is insufficient for assessing the structural behaviour and the stability of structures in/on geo-materials having minerals prone to water absorption/desorption. An appropriate model definitely requires the information on the moisture migration characteristics of geo-materials and the variations of their physical and mechanical properties with the water content.

This chapter first describes a theoretical method to model the water-content migration in geomaterials. Then some experimental set-ups are presented to measure the moisture migration and associated volumetric variations of geo-materials prone to water absorption/desorption. Then, physical and mechanical of properties of soft rocks are measured in relation to water content.

3.2 MODELING OF WATER ABSORPTION/DESORPTION PROCESSES AND ASSOCIATED VOLUMETRIC CHANGES IN ROCKS

Some rocks such as fine grain sandstone, mudstone and siltstone start to fracture during losing their water content as is observed in many laboratory tests and in-situ. The situation is similar to reverse problem of swelling problem. It is considered that rock shrinks as it loses its water content. This consequently induces results in shrinkage strain leading to fracturing of rock in tension. Therefore, a coupled formulation of the problems is necessary.

3.2.1 Mechanical modeling

The water content variation in rock can be modeled as a diffusion problem. Thus the governing equation is written as

$$\frac{d\theta}{dt} = -\nabla \cdot \mathbf{q} + Q \tag{3.1}$$

where θ, $\mathbf{q}$, Q and t are water content, water content flux, water content source and time, respectively. If water content migration obeys Fick's law, the relation between flux $\mathbf{q}$ and water content is written in the following form:

$$\mathbf{q} = -k\nabla\theta \tag{3.2}$$

where k is water diffusion coefficient. If some water content is transported by the ground water seepage or airflow in open space, this may be taken into account through the material derivative operator in Eq. (3.1). However, it would be necessary to describe or evaluate the seepage velocity or airflow.

If the stress variations occur at slow rates, the equation of motion without inertial term may be used in incremental form as given below:

$$\nabla \cdot \dot{\sigma} = 0 \tag{3.3}$$

The simplest constitutive law for rock between stress and strain fields would be a linear law, in which the properties of rocks may be related to the water content in the following form (i.e. Aydan *et al.*, 2004):

$$\dot{\sigma} = \mathbf{D}(\theta)\dot{\varepsilon}_e \tag{3.4}$$

The volumetric strain variations associated with shrinkage (inversely swelling) may be related to the strain field in the following form:

$$\dot{\varepsilon}_e = \dot{\varepsilon} - \dot{\varepsilon}_s \tag{3.5}$$

3.2.2   Finite element modeling

The finite element form of water content migration takes the following form after some manipulations of Eq. (3.1) and Eq. (3.2) through usual finite element procedures:

$$[M]\{\dot{\theta}\} + [H]\{\theta\} = \{Q\} \tag{3.6}$$

where

$$[M] = \int [N]^T [N] dV; \quad [H] = k \int [B]^T [B] dV; \quad \{Q\} = \int [\overline{N}]^T \{q_n\} d\Gamma$$

Similarly, the finite element form of incremental equation of motion given by Eq. (3.6) is obtained as follows:

$$[K]\{\dot{U}\} = \{\dot{F}\}$$

where

$$[K] = \int_V [B]^T [D][B] dV; \quad \{\dot{F}\} = \int_V [B]^T [D]\{\dot{\varepsilon}_s\} dV + \int_S [\overline{N}]^T \{\dot{t}\} dS$$

3.3   MOISTURE MIGRATION PROCESS AND VOLUMETRIC CHANGES

Water migration in geomaterials takes place in two different forms, namely, molecular diffusion and seepage. The seepage phenomenon involves the relative motion of water with respect to solid phase and water is free to move within the solid skeleton if any pressure gradient exists. Molecular diffusion is an interaction between water molecules and solid phase and water is either absorbed to or desorbed from the solid phase. This section is concerned with the development of a theoretical model for determining the water content migration properties of geomaterials during drying or saturation tests.

The details of this method are described and the results of its applications to actual tests are presented in this section.

3.3.1 Drying testing procedure

Let us consider a sample with volume V dried in air with infinite volume as shown in Figure 3.2 (Aydan, 2003a). Water contained Q in a geo-material sample may be given in the following form

$$Q = \rho_w \theta_w V \qquad (3.7)$$

where ρ_w, θ_w and V are water density, water content ratio and volume of sample, respectively. Assuming that water density and sample volume remain constant, the flux q of water content may be written in the following form

$$q = \frac{dQ}{dt} = -\rho_w V \frac{d\theta_w}{dt} \qquad (3.8)$$

Air is known to contain water molecules of 6 g/m³ when relative humidity is 100%. When the relative humidity is less than 100%, water is lost from geomaterials to air. If such a situation presents, the water lost from the sample to air may be given in the following form using a concept similar to Newton's cooling law in thermo-dynamics:

$$q = \rho_w A_s h \Delta \theta = \rho_w A_s h (\theta_w - \theta_a) \qquad (3.9)$$

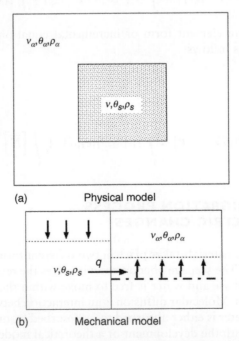

(a)   Physical model

(b)   Mechanical model

Figure 3.2 Physical and mechanical models for water migration during drying process.

where h and A_s are water loss coefficient and surface area of sample. Requiring that the water loss rate of sample should be equal to the water loss into air on the basis of the mass conservation law, one can easily write the following relation

$$\rho_w A_s h(\theta_w - \theta_a) = -\rho_w V \frac{d\theta_w}{dt} \tag{3.10}$$

The solution of differential equation (3.10) is easily obtained in the following form

$$\theta_w = \theta_a + Ce^{-\alpha t} \tag{3.11}$$

where

$$\alpha = h\frac{A_s}{V}$$

The integration constant may be obtained from the initial condition, that is,

$$\theta_w = \theta_{w0} \quad \text{at } t = 0 \tag{3.12}$$

as follows

$$C = \theta_{w0} - \theta_a \tag{3.13}$$

Thus the final expression takes the following form

$$\theta_w = \theta_a + (\theta_{w0} - \theta_a)e^{-\alpha t} \tag{3.14}$$

If the water content migration is considered as a diffusion process, Fick's law in one dimension may be written as follows:

$$q = \rho_w D \frac{\partial \theta_w}{\partial x} \tag{3.15}$$

Requiring that water loss rate given by Eq. (3.15) to be equal to that given by Eq. (3.9) yield the following relation

$$D = h\frac{V}{A_s} \tag{3.16}$$

If surface area A_s and volume V of sample are known, it is easy to determine the water migration diffusion constant D from drying tests easily, provided that the coefficient α and subsequently h are determined from experimental results fitted to Eq. (3.14).

If samples behave linearly, water migration characteristics should remain the same during swelling and drying processes. Recent technological developments have made it quite easy to measure the weight of samples and the environmental conditions such as temperature and humidity. Figure 3.3 shows an automatic weight and environmental

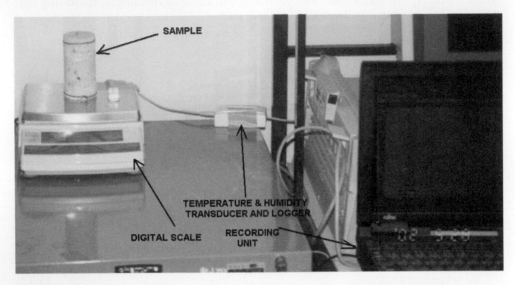

Figure 3.3 The experimental set-up for measuring water content during drying.

Table 3.1 XRD results from the samples of Ürgüp (Kavak tuff) and Avanos.

Specimen Number	Clay Percentage	Clay fraction		
		Smectite	Kaolin	Illite
UR-1 (Ürgüp)	74	83	14	3
UR-2 (Ürgüp)	60	67	25	8
AV-1 (Avanos)	94	84	13	3
AV-2 (Avanos)	82	95	5	T

T: Trace amount

conditions monitoring system developed for such tests. It is also possible to measure the volumetric variations (shrinkage) during drying process using non-contact type displacement transducers (i.e. laser transducers).

Physical and mechanical properties of materials can be measured using the conventional testing machines such as wave velocity measurements, uniaxial compression tests, elastic modulus. Tuff samples used in the tests were from Avanos, Ürgüp and Derinkuyu of Cappadocia Region in Turkey and Oya in Japan. The samples from Cappadocia region are gathered from historical and modern underground rock structures. They represent the rocks in which historical and modern underground structures were excavated. These tuff samples bear various clay minerals as given in Table 3.1 (Temel, 2002; Aydan & Ulusay, 2003). As noted from the table, the clay content is quite high in Avanos tuff and most of the clay minerals are smectite.

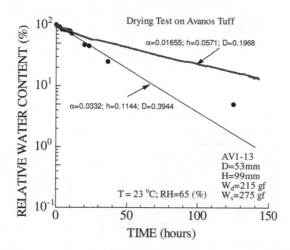

Figure 3.4 Determination of constants for relative water content variation during drying of Avanos tuff.

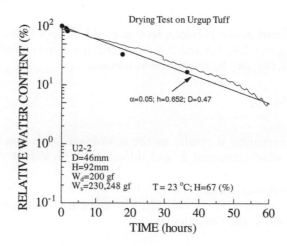

Figure 3.5 Determination of constants for Relative water content variation during drying of Ürgüp tuff.

In drying experiments, the samples underwent swelling were dried in room with an average temperature of 23°C and relative humidity of 65–70. Figures 3.4, 3.5 and 3.6 show the drying test results for some tuff samples from Cappadocia region in Turkey. As seen from the figures, it takes a longer time for the tuff sample from Avanos compared with Ürgüp and Derinkuyu samples. Derinkuyu sample dries much rapidly than the others. Each sample was subject to drying twice. Once again it is noted that the drying period increases for Avanos tuff after each run while Derinkuyu tuff tends to dry much rapidly in the second run. From these tests, it may be also possible to determine the diffusion characteristics of each tuff.

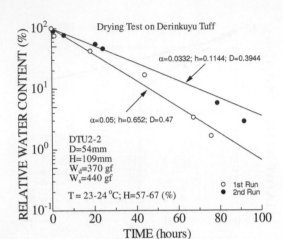

Figure 3.6 Determination of constants for Relative water content variation during drying of Derinkuyu tuff.

The theory derived in the previous section could be applied to the experimental results shown in Figures 3.4, 3.5 and 3.6. To obtain the constants of water migration model, Equation (3.14) may be re-written as follows:

$$\ln\left(\frac{\theta_w - \theta_a}{\theta_{w0} - \theta_a}\right) = -\alpha t \tag{3.17}$$

The plot of experimental results in the semi-logarithmic space first yields the constant α, from which constant h and diffusion coefficient D can be computed subsequently.

The results are shown in Figures 3.4, 3.5 and 3.6. The unit of parameters α, h and D are 1/hr, cm/hr and cm^2/hr, respectively. The computed values of parameters α, h and D are also shown in the same figures.

3.3.2   Saturation testing technique

Initially dry samples can be subjected to saturation and water migration characteristics may be obtained. The side of samples can be sealed and subjected to saturation from the bottom. The top surface may be sealed and unsealed as illustrated in Figure 3.7. Samples can be isolated against water migration from sides by sealing while the bottom surface of the samples can be exposed to saturation by immersing in water up to a given depth. There may be two conditions at the top surface, which could be either exposed to air directly or sealed. When the top surface is sealed, the boundary value would be changing with time. The water migration coefficient can be determined from the solution of the following the diffusion equation:

$$\frac{\partial \theta_w}{\partial t} = D \frac{\partial^2 \theta_w}{\partial x^2} \tag{3.18}$$

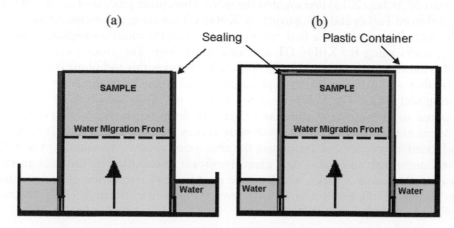

Figure 3.7 Experimental set-ups: (a) top surface unsealed; (b) Top-surface sealed.

When the top surface is unsealed, the top boundary condition $(x = H)$ is

$$\theta_w = \theta_a. \tag{3.19}$$

On the other hand, if the top surface is sealed, the boundary condition is time dependent and it can be estimated from the following condition

$$q_{x=H} = \hat{q}_n(t) \tag{3.20}$$

For some simple boundary conditions, the solution of partial differential Eq. (3.19) can be easily obtained using the technique of separation of variables (i.e. Keryzig, 2011). In general case, it would be appropriate to solve it using finite difference technique or finite element method (i.e. Aydan 2003, Section 3.2).

3.3.3 X-Ray Computed Tomography (CT) scanning technique

X-Ray Computed Tomography (CT) imaging technique can be used to visualize water-absorption process of soft rocks samples. Furthermore, it is expected that such evaluation may reveal the basic mechanism of water absorption or desorption in soft rocks (Sato & Aydan, 2013).

X-Ray CT scanners can be used to investigate the internal structure of materials and various processes without disturbance to samples (Ketcham & Carlson, 2001). The device was a μ-focus X-ray CT Scanner System (TOSCANER-32300FPD) operated by Kumamoto University (i.e. Sato *et al.*, 2011a,b). This system has relatively higher X-ray emission (230 keV/608 μA) and the minimum 4 μm focus distance is available. X-ray CT scanner is the system to detect the density distribution in materials. However, large density difference in a tomographic region cannot be correctly visualized. This technique was used to investigate water absorption and diffusion processes in soft rocks.

Sato & Aydan (2013) investigated the water absorption process of some soft rocks sampled from Turkey and Japan using the X-Ray CT scanning technique. As mentioned in the introduction, this is a first attempt to evaluate the moisture migration process in soft rocks using the X-Ray CT scanning technology. The process can be investigated without any disturbance to samples, which makes this technique quite suitable to visualize and quantify the absorption process of water by minerals or grains constituting soft rocks. The water migration process in Cappadocia tuffs is quite rapid compared to that of Asuwayama and Oya tuffs from Japan. Bazda limestone from the Bazda antique quarries from south-west Turkey was also quite rapid. If numerical simulations are carried out to simulate the absorption process visualized by the X-Ray CT technique, this could also yield great significant information on the mechanism of degradation process of rocks, which can be used for the preservation of antique structures as well as the assessment of long term stability of rock engineering structures involving soft rocks.

(a) Experimental Set-up

The boundary condition shown in Figure 3.7(a) was used to investigate the water absorption process. Samples were isolated against water migration from sides by sealing and the bottom and top surfaces of the samples were exposed to water and air, respectively. First CT scanning of dry samples was carried out. Then, samples exposed to water from the bottom, and CT scanning of samples exposed to water migration was measured at certain time intervals. The weight of samples was measured at each time step and CT images were used to evaluate the water content variation in samples. The differences in CT values of samples under dry state and exposed to water migration at a given time step were used to evaluate the water content variation of samples. CT value differences ranged between 200 and 300. The ratio of CT value of water to that of air is about 1000. Therefore, the measured CT value difference implies that the specific density of samples increased by 0.2 to 0.3 times.

(b) Results and Discussions

The water absorption processes in five different soft rocks are measured using the X-Ray CT scanning technique are described and results are discussed. Results are presented as X-Ray CT scanning images and variation of CT values with height.

Asuwayama Tuff

Asuwayama tuff has been extracted from the Asuwayama hill in Fukui City and it is commercially known as Shakutani stone. The UCS of this tuff is about 30 MPa and it is dense. Figure 3.8 shows X-Ray CT scanning images and CT value distribution with height at time intervals of 1 hr, 2, 3 and 4 hrs. As expected, water migrates upward gradually. The water absorption front is clearly observed in CT scan images.

Oya Tuff

Oya tuff has been extracted from the Oya town of Utsunomiya City and it is commercially known as Oya stone. The UCS of this tuff is about 10 MPa. Figure 3 shows X-Ray CT scan images and CT value distribution with height at time intervals of

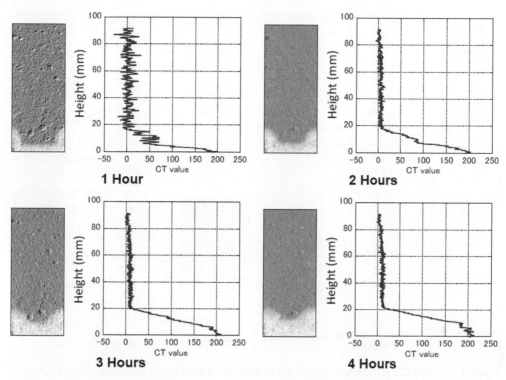

1 Hour

2 Hours

3 Hours

4 Hours

Figure 3.8 X-Ray CT scan images and CT value distribution with height at different time intervals in Asuwayama tuff sample.

1 hr, 2, 3 and 4 hrs. As expected, water migrates upward gradually. Although the water absorption front is clearly observed in CT scanning images, the front is not straight. This may be due to inclusions of highly absorptive nodules such as clays in the sample.

Bazda Limestone

Bazda limestone has been extracted from Bazda antique quarries, which may be more than 3000 years old, in the Tektek Mountains in the southwest of Turkey. The UCS of this limestone is about 15 MPa and it is relatively dense. Figure 3.10 shows X-Ray CT scan images and CT value distribution with height at time intervals of 1 hr, 2, 3 and 4 hrs. As expected, water migrates upward gradually and the water absorption front migrates much more quickly compared to those of Oya and Asuwayama tuffs. The water absorption front is clearly seen in CT scanning images.

Zelve Tuff

Zelve tuff is found in the Zelve Valley of Cappadocia region of Turkey. There are many antique settlements in this valley, which are more than 1500 years old at least. The UCS of this tuff is about 4–5 MPa under dry state. Figure 3.11 shows X-Ray CT

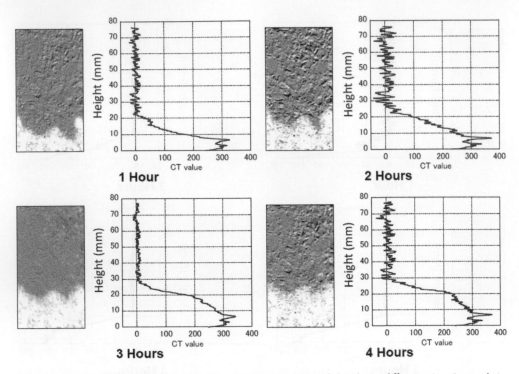

Figure 3.9 X-Ray CT scan images and CT value distribution with height at different time intervals in Oya tuff sample.

scan images and CT value distribution with height at time intervals of 1 hr, 2, 3 and 4 hrs. As expected, water migrates upward gradually and the water absorption front migrates much more quickly compared to those of Oya and Asuwayama tuffs. The water absorption front is clearly seen in CT scanning images.

3.4 SWELLING-SHRINKAGE PROCESS

3.4.1 Shrinkage process

Next, the water content migration characteristics and associated volumetric variations were measured. For this purpose, an experimental device illustrated in Figure 3.12 was used (Aydan *et al.*, 2006). The experimental set-up consists of an automatic scale, an electric current inductor, electrodes, isolators, laser displacement transducer, voltmeter, rock sample, lap-top computers to monitor and to store the measured parameters and temperature-humidity unit consisting of sensors and logger. Rock samples were first fully soaked with water for a certain period of time. Then they were put on the automatic scale and dried. During the drying process, the weight, length and voltage changes of the sample were continuously measured. The temperature and humidity changes of the drying place were also continuously monitored.

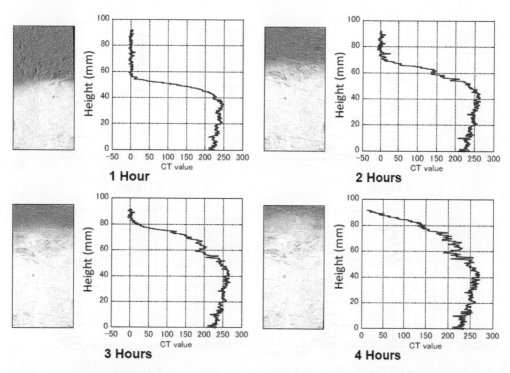

Figure 3.10 X-Ray CT scan images and CT value distribution with height at different time intervals in Bazda limestone sample.

Figure 3.13 shows temperature, humidity, shrinkage strain, weight change and electrical resistivity variations on both fine grain sandstone and coarse grain sandstone samples in laboratory. While the weight change (water content) of coarse grain sandstone was slightly larger than that of fine-grain sandstone, there was a remarkable difference between the shrinkage strains of samples. The shrinkage strain of fine grain sandstone was more than twice that of coarse grain sandstone.

Aydan *et al.* (2006) reported that the electrical resistivity of samples increases as the samples lose their water content. It is considered that if the electrical resistivity of surrounding rock could be measured continuously in-situ, it may be quite useful for evaluating the water content variations and associated volumetric variations.

3.4.2 Swelling process

Swelling minerals in rocks may be broadly classified into the following groups:

- *Clay minerals (montmorillonite, bentonite, smectite, corensite):* Among the clay minerals, montmorillonite exhibits the largest swelling potential. Under unstressed state, the volumetric expansion varies from 40 to 200% (Özkol, 1965; Brekke, 1965; Pasamehmetoglu *et al.*, 1993; Yesil *et al.*, 1993).

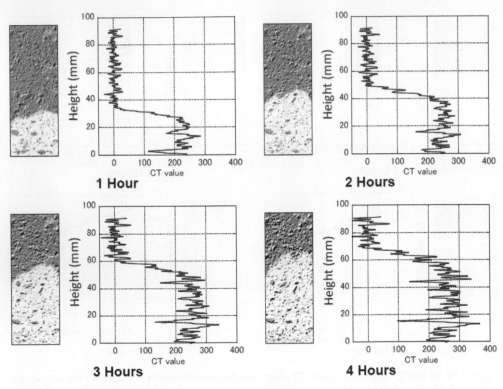

Figure 3.11 X-Ray CT scan images and CT value distribution with height at different time intervals in Zelve tuff sample.

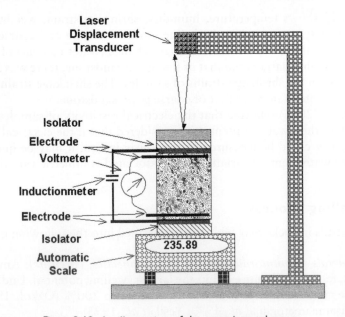

Figure 3.12 An illustration of the experimental set-up.

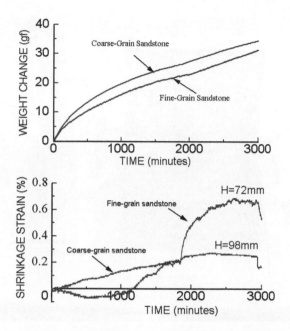

Figure 3.13 Water-content migration tests and associated volumetric changes in laboratory.

- Evaporitic minerals (anhydrite, gypsum): The transformation of anhydrite into gypsum results in a 60% volumetric expansion when it adsorbs water under unstressed state. Volumetric decrease occurs when gypsum transforms into anhydrite as it losses its water content (Vardar & Fecker 1986; Zanbak & Arthur 1985).

The swelling potentials of minerals and rocks have been tested by several researchers (i.e. Özkol, 1965; Brekke, 1965; Murayama & Yagi, 1966; Marsden *et al.*, 1992). The swelling potential of rocks commonly assumed to be a function of a period of saturation and compressive volumetric stress (Wittke, 1990; Franklin & Dusseault, 1989). However, the swelling potential must be related to the water content instead of the period of saturation since the swelling depends upon the water content.

Aydan *et al.* (1993, 1994) undertook an experimental study to determine the swelling potential of geo-materials proposed to model the mechanical effect of swelling process of geo-materials in engineering problems. An experimental device developed by Aydan (2003a) is used to determine the parameters of the swelling potential function given below (Figures 3.14 and 3.15):

$$\varepsilon_{sv} = f(\theta, \sigma_v) \tag{3.21}$$

where ε_{sv}: swelling strain; σ_v: volumetric stress (or pressure), θ: water content.

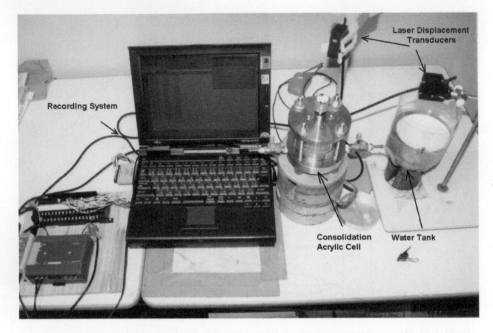

Figure 3.14 Experimental set-up for measuring swelling and water content.

The swelling potential function above may be determined from a combination of free-swelling test and compression test on a single sample, if the swelling potential function is specifically chosen as given below:

$$\varepsilon_{sv} = A(1 - e^{-B\theta})e^{-C\sigma_v} \qquad (3.22)$$

where A, B, C are constants to be determined from tests. Since volumetric stress σ_v is nil during free-swelling tests, constants A and B can be easily determined from the measured swelling strain and water content response together with a simple curve-fitting procedure. By keeping the water content constant and applying compression onto the specimen, constant C can be obtained from the volumetric stress and strain response with the utilization of the curve-fitting procedure.

Displacement of specimens during the processes of free swelling and compression was measured with laser displacement transducers. Furthermore, the water content during the free-swelling process was measured through the variation of the level of the water supply tank as illustrated in Figures 3.14 and 3.15a. In this procedure, one should take care of minimizing the water loss from the system as a result of evaporation. If the air of the testing environment is kept at the relative humidity of 100%, there will be almost no water loss due to the evaporation.

In swelling experiments, the samples of bentonitic clay, which were initially oven-dried, were fully submerged in tap water with a pH value of 7.6–7.8. The samples were sealed in a manner so that no water-loss occurs during the swelling and compression processes. Figures 3.16 and 3.17 show the swelling and compression responses measured during one of the experiments on a bentonitic clay buffer material.

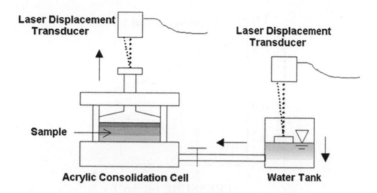

(a) Free swelling set-up

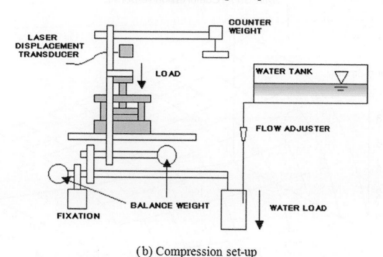

(b) Compression set-up

Figure 3.15 Drawing of the experimental setup shown in Figure 3.14.

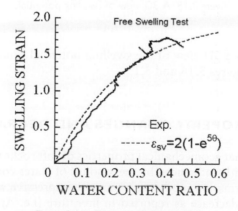

Figure 3.16 Free-swelling response.

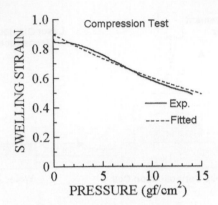

Figure 3.17 Compression response.

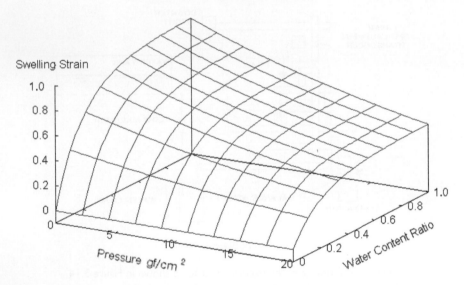

Figure 3.18 A 3D view of swelling potential.

Figure 3.18 shows a 3D view of the swelling potential function determined from responses shown in Figures 3.16 and 3.17.

3.5 MATERIAL PROPERTY CHANGES AND DEGRADATION

It is very well known that geomechanical properties of soft rocks with water absorption characteristics are greatly influenced by the amount of water content. It is experimentally shown that deformation modulus, uniaxial compressive and tensile strength of soft rocks drastically decrease as reported in literature (i.e. Aydan, 2003; Aydan & Ulusay 2003, 2013). It is also reported that even the geomechanical properties of

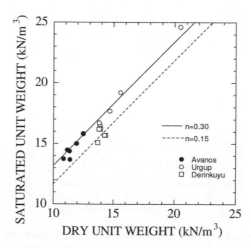

Figure 3.19 Relation between dry unit weight and saturated unit weight.

very hard rocks may be influenced by the water content (i.e. Colback & Wiid, 1965; Broch, 1979).

It is well known that the water content in rocks may influence the mechanical properties of rocks. The effect of water content on some physical and mechanical properties of Cappadocia and Oya tuffs is briefly described below.

The unit weight of rocks differs as its water content increases. Figure 3.19 shows the relation between the dry unit weight and fully saturated unit weight of tuffs. This relation may be approximated through the following relation based on the mixture theory (i.e. Aydan 1992; Aydan *et al.*, 1996):

$$\gamma_s = \gamma_d + n\gamma_w \tag{3.23}$$

where γ_s, γ_d, γ_w and n are the unit weight of saturated and dry samples and water, and volume fraction, respectively. In the figure, two lines are drawn with different volume fraction ratios. Since the volume fraction of the samples varied between 0.13 and 0.30, the fitted lines should be relevant to the experimental results.

The elastic wave velocity of rocks differs as its water content increases. The relations shown in Figure 3.20 can be obtained through the use of the mixture theory together with parallel and series model concepts (i.e. Aydan, 1992; Aydan *et al.*, 1996):

Parallel Model

$$\frac{V_{pm}}{V_{pd}} = 1 + n \cdot S \frac{V_{pw}}{V_d} \tag{3.24}$$

Series Model

$$\frac{V_{pm}}{V_d} = \frac{V_{pw}/V_{pd}}{nS + V_{pw}/V_{pd}} \tag{3.25}$$

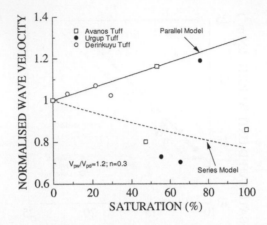

Figure 3.20 Relation between saturation and wave velocity.

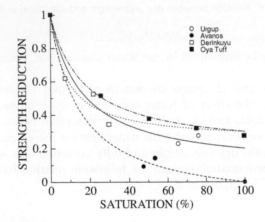

Figure 3.21 Relation between saturation and uniaxial compressive strength.

where V_{pm}, V_{pd}, V_{pw} and S are the wave velocity of saturated and dry samples and water, and saturation coefficient, respectively. In Figure 3.20 these two lines are drawn for $n = 0.3$ and $V_{pw}/V_{pd} = 1.2$. The comparison of experimental results with theoretical predictions indicates that both models may be appropriate.

Figure 3.21 shows the relation between the saturation (S) and normalized compressive strength of Cappadocia tuffs by their uniaxial compressive strength at dry state. The lines drawn in the figure are obtained from the following empirical relation:

$$\frac{\sigma_{cw}}{\sigma_{cd}} = \alpha_o - (\alpha_0 - \alpha_{100})\frac{S}{S + \beta(100 - S)} \tag{3.26}$$

where $\sigma_{cw}, \sigma_{cd}, \alpha_0, \alpha_{100}$ and β are uniaxial compressive strength of sample containing water, uniaxial compressive strength at dry state, empirical coefficients, respectively.

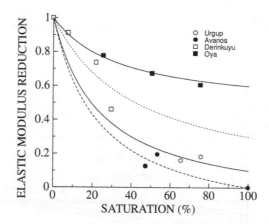

Figure 3.22 Relation between saturation and elastic modulus.

In the same figure, the results for Oya tuff, which is a well known soft rock in Japan are also plotted.

The elastic modulus of rocks differs as its water content increases. Figure 3.22 shows the relation between the saturation and normalized elastic modulus of Cappadocia tuffs by their elastic modulus at dry state. The lines drawn in the figure are obtained from Equation 3.27 for E_w/E_d which are

$$\frac{E_w}{E_d} = \eta_o - (\eta_0 - \eta_{100})\frac{S}{S + \theta(100 - S)} \qquad (3.27)$$

where $E_w, E_d, \eta_0, \eta_{100}$ and θ are elastic modulus of sample containing water and elastic modulus at dry state. The variations of physical and mechanical properties of the Avanos tuff are larger than those for the tuffs from Oya, Ürgüp and Derinkuyu.

3.6 APPLICATIONS

3.6.1 Breakout formation in rocks due to moisture loss

The first analyses were concerned with the simulations of displacement, strain and stress field around a circular borehole in a hydrostatic stress field. Specifically, the effects of sandstone type and diameter of borehole were analyzed. Figure 3.23 shows the computed results for displacement, water content and stress fields for fine and coarse grain sandstones for a borehole with a diameter of 200 mm under the overburden of the adit at several time steps.

As the water migration characteristics of both fine and coarse sandstones were same, the resulting water content migration distributions with time were same. However, displacement, strain and stress fields were entirely different for each sandstone type. As the volumetric variation of fine grain sandstone as a function of water content is much larger than that of coarse grain sandstone, the shrinkage of the borehole in

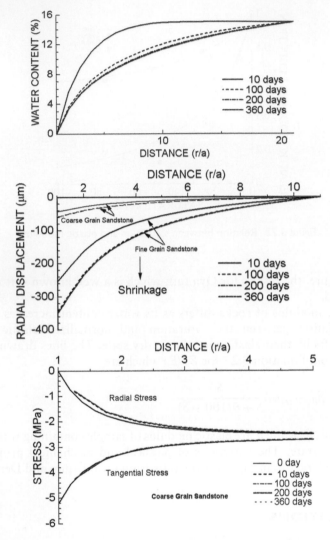

Figure 3.23 Variations of computed water content, displacement, stress fields for fine and coarse grain sandstones.

fine grain sandstone is larger than that in coarse grain sandstone. Consequently, radial stress in the close vicinity of the borehole wall becomes tensile in fine grain sandstone. This, in turn, implies that there would be fractures parallel to the borehole wall if the tensile strength of rock were exceeded. Furthermore, such fractures would only occur in the vicinity of boreholes in fine grain sandstone, as it is observationally noted in-situ.

The next computational example was concerned with a circular borehole under two dimensional in-situ stress fields as shown in Figure 3.24. It is observed that the bottom of the borehole was wet or covered with water in-situ. In order to take into account this observation in computations, the boundary conditions for water content migration field and displacement field were assumed as illustrated in Figure 3.24.

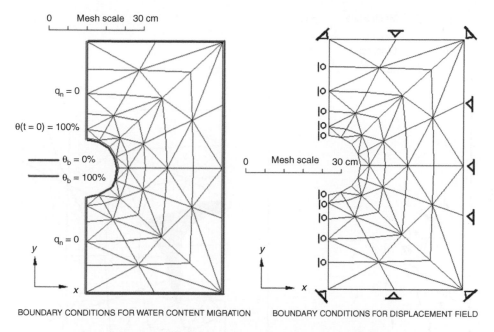

BOUNDARY CONDITIONS FOR WATER CONTENT MIGRATION   BOUNDARY CONDITIONS FOR DISPLACEMENT FIELD

Figure 3.24 Assumed boundary conditions in computations.

The other properties were the same as those used in axisymmetric simulations. The computed displacement field and associated yielding zone are shown in Figure 3.25.

As noted from the figure, the bottom of the borehole heaves and crown of the borehole shrinks upward. In other words, the upper part of the borehole expands outward due to water content loss. The displacement and stress fields of surrounding rock are entirely different at the lower and upper parts of the borehole. As a result of this fact, yielding occur only in the upper part of the borehole. This computational result is in accordance with actual observations. The yielding zone is not depleted in this computation. However, if the yielding zone were depleted in the computation region, the process would repeat itself after depletion of the yielded zone each time.

3.6.2 Tunneling in swelling rocks

The example is associated with the simulations of tunnel excavations in swelling rocks. A circular tunnel of 4 m in diameter was assumed to be situated in a hydrostatic state of stress initially. Table 3.2 gives the properties of surrounding ground and constants of swelling potential function. The deformation responses of tunnels in swelling rocks are generally influenced by the following factors:

- Installation timing of lining,
- Rigidity of lining,
- Depth of tunnel, and
- Properties of lining material.

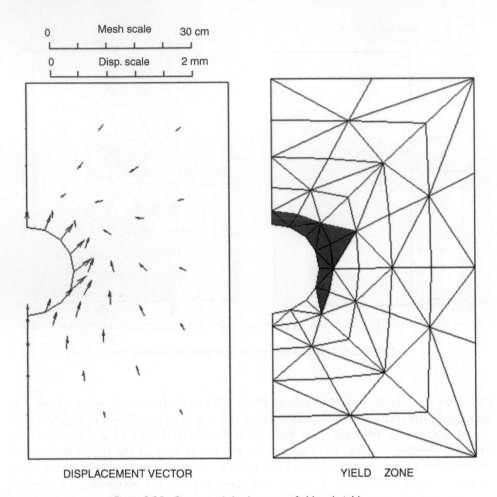

Figure 3.25 Computed displacement field and yield zone.

A series of parametric studies was carried out to check the influences of the some of the factors above on the water content, deformation of surrounding rock and tangential stress in lining. In the following presentation, the results on the tangential stress of the lining are only reported and discussed. By assuming that the initial in-situ stress was 1.0 MPa, Figure 3.26 shows the tangential stress variation of the lining as a function of time. The delay periods were set as 10 days, 20 days and 30 days. As the period of delay increases, the lining stress becomes larger. This implies that the variation of water content in rock must be prevented as quickly as possible. Otherwise, the cracking of the lining is unavoidable since the lining stress increases with the delay.

To see the effect of tunnel depth, the initial in-situ stress was varied from 1.00 MPa to 1.25 MPa. The results of calculations are shown in Fig. 3.27. The volumetric stress decreases as the tunnel becomes shallow. Consequently, if there is any water content

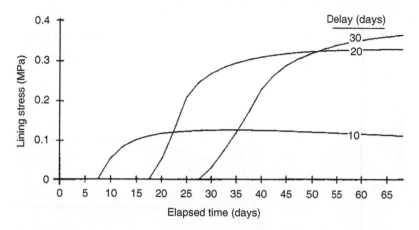

Figure 3.26 Stress acting on tunnel lining for several delay time of installation.

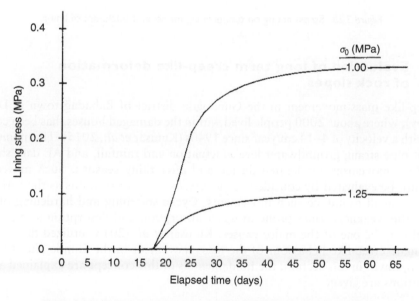

Figure 3.27 Stress acting on tunnel lining for several in-situ condition.

variation, a large deformation of surrounding ground occurs. As a result of that, the tangential stress in the lining becomes larger as the depth becomes shallow as shown in Figure 3.27.

Figure 3.28 shows the tangential stress variation of the lining as a function of its thickness. Since the rigidity of lining increases as a result of larger thickness, the resulting tangential stress in the lining becomes smaller. This example is qualitatively in good agreement with observations in tunnelling in swelling ground.

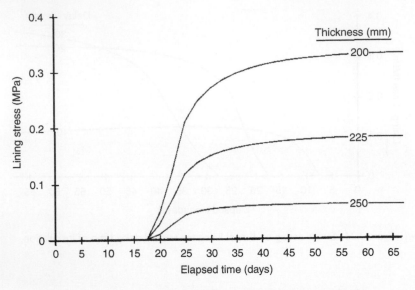

Figure 3.28 Stress acting on tunnel lining for several thickness of lining.

3.6.3 Evaluation of long term creep-like deformation of rock slopes

A creep-like mass movement in the Gündoğdu district of Babadağ town in Denizli (Turkey), where about 2000 people lived within the damaged houses, has been continuing with a velocity of 4–14 cm/year since 1940s (Kumsar *et al.*, 2015). The monitoring data of pipe strain, groundwater level fluctuation and rainfall, and AE data showed that slope movement accelerated during and after rainy seasons. Such movements could not be evaluated by considering the effective stress variations as the variations are quite small to induce such movements. Cyclic softening and hardening of stiffness of the weakness zones prone to water absorption and desorption as a result of rainfall may be one of the major causes. Kumsar *et al.* (2015) utilized that concept to evaluate creep-like behaviour of the mass movement in the Gündoğdu district of Babadağ town. In this section, the fundamentals of this concept are explained and its applications are given.

3.6.3.1 Analytical model and its application

The simplified analytical model introduced in this section is based the theoretical model developed by Aydan (1994, 1998). Momentum conservation law for infinitely small element of a ground on a plane with an inclination of α for each respective direction can be written in the following form (Figure 3.29)

x-direction

$$\frac{\partial \tau}{\partial y} = \frac{\partial p}{\partial x} - \rho g \sin \alpha \tag{3.28}$$

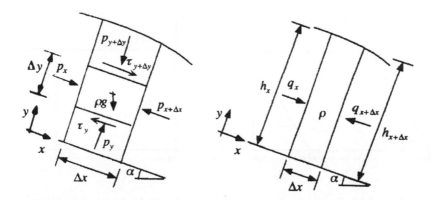

Figure 3.29 Modelling of a layer subjected to shearing (from Aydan 1994, 1998).

y-direction

$$\frac{\partial p}{\partial y} = \rho g \sin \alpha \tag{3.29}$$

where τ, p, ρ, g are shear stress, pressure, density and gravitational acceleration, respectively. The variation of pressure along x-direction is given by

$$\frac{\partial p}{\partial x} = \rho g \cos \alpha \frac{\partial h}{\partial x} \tag{3.30}$$

If shear stress related to shear strain is linearly as given in the following form

$$\tau = G\gamma; \quad \gamma = \frac{\partial u}{\partial y}$$

one can easily obtain the solution given as

$$\tau = \rho g \cos \alpha \left(\tan \alpha - \frac{\partial h}{\partial x} \right) (h - y) \tag{3.31}$$

If the variation of ground surface height (h) is neglected, the resulting equation for shear stress and displacement takes the following form

$$\tau = \rho g \sin \alpha (h - y); \quad u = \frac{\rho g \sin \alpha}{G} y \left(h - \frac{y}{2} \right) \tag{3.32}$$

As well known, the rainfall induces groundwater level fluctuations. However, these fluctuations are not that high as presumed in many limiting equilibrium approaches to analyze the failure of slopes. In other words, the whole body, which is prone to fail, do not become fully saturated. However, the monitoring results indicate that a certain thickness of layer becomes saturated. In view of experimental results, the

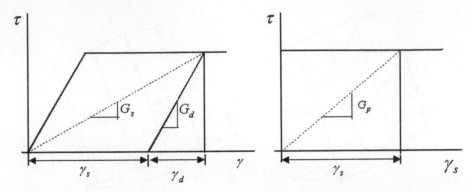

Figure 3.30 Constitutive modelling of cyclic softening-hardening of marl layer.

deformation modulus would become smaller during saturation process and recover its original value upon drying. The deformation modulus during saturation may be assumed to be plastic deformation modulus (G_p) and the displacement induced during saturation period may be viewed as plastic (irrecoverable) deformation (Figure 3.30). With the use of this concept and the analytical model presented above, one can easily derive the following equation for deformation induced by saturation as

$$u_s = \frac{\rho g \sin \alpha}{G_s} y \left(h - \left(t - \frac{y}{2} \right) \right) \tag{3.33}$$

where t is the thickness of saturated zone in a given cycle of saturation-drying. The plastic deformation would be the difference between displacements induced under saturated and dry states and it will take the following form:

$$u_p = \rho g \sin \alpha y \left(\frac{1}{G_s} - \frac{1}{G_d} \right) \cdot \left(h - \left(t - \frac{y}{2} \right) \right) \tag{3.34}$$

where G_d and G_s are shear modulus for dry and saturated states, respectively. Thus the equivalent shear modulus can be written as and it is called as plastic deformation modulus (G_p) in this study

$$G_p = \frac{G_s G_d}{G_d - G_s} \tag{3.35}$$

The time for saturation and drying of marls is very short (say, in hours). With this observational fact and experimental results, the analysis presented is based the day unit. Figure 3.31 compares the computed displacement and displacement measured at monitoring station No. 1 the Gündoğdu district of Babadağ town with the consideration of thickness of the saturation zone. Despite some differences between computed and measured responses, the analytical model can efficiently explain the overall response of the landslide area of Gündoğdu district of Babadağ town.

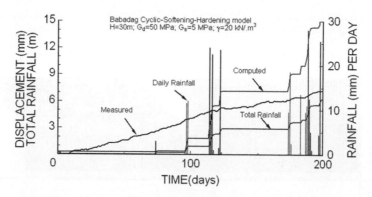

Figure 3.31 Comparison of measured and computed displacements.

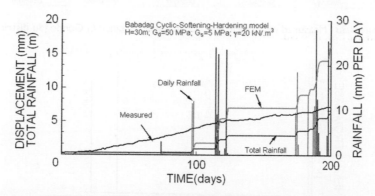

Figure 3.32 Comparison of measured and computed displacements.

3.6.3.2 *Semi-infinite multi-layer finite element model and its application*

If the variation of thickness of the saturation zone is given, a finite element version of analytical model given by Eqs. (3.26, 3.29) can be easily developed. The finite element formulation of Eq. (3.28) may be written as:

$$[K]\{U\} = \{F\} \tag{3.36}$$

where

$$[K] = \int_{y_i}^{y_j} [B]^T G(\theta)[B] dy; \quad \{F\} = -\rho g \sin \alpha \int_{y_i}^{y_j} [N]^T dy \tag{3.37}$$

As shear modulus $G(\theta)$ depends upon saturation, non-linear analysis is necessary. To deal with this non-linearity, the behaviour of saturated layer would be quite similar to elastic-perfectly plastic materials as illustrated in Figure 3.30. During the solution in time-space, the thickness of the layer changes depending upon the amount of rainfall. Figure 3.32 shows the computational results for the situation analysed in Figure 3.31.

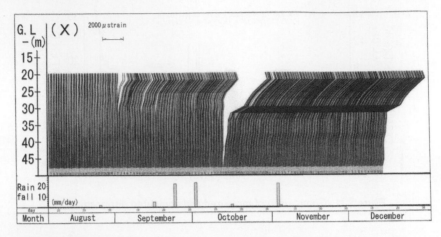

Figure 3.33 Measured results for pipe-strain meter installed in Gündoğdu district.

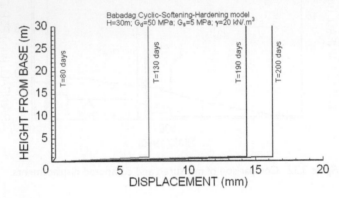

Figure 3.34 Computed displacement responses at selected time steps.

When computed results from analytical model and the finite element model are compared with each other, one can easily notice that the results are almost the same. The slight difference is related to the small error caused from the numerical discretization by the finite element method.

Figure 3.33 shows the measured results for pipe-strain meter installed in Gündoğdu district (Kumsar *et al.*, 2015). As noted from the figure strain become larger at certain depth. Figure 3.34 shows the horizontal displacement response above the fixed base. It is interesting to notice that the overall behaviour from the finite element analysis resembles to that measured by the pipe-strain meter. In other words, this model presented herein can clearly evaluate the creep-like deformation of the Gündoğdu district, which could not be evaluated in classical sliding type analyses.

3.6.3.3 Implementation in discrete finite element method (DFEM) and analyses

Aydan and his co-workers (Aydan *et al.*, 1996; Aydan & Mamaghani 1996; Mamaghani 1994, 1995, 1996) developed a finite element model to handle large deformations resulting from rigid-body like motions due to sliding or separation and they named this method as discrete finite element method (DFEM). It consists of a mechanical model to represent the deformable blocks and contact models that specify the interaction among them. Small displacement theory is applied to the intact blocks while blocks can take finite displacement. Blocks are polygons with an arbitrary number of sides, which are in contact with the neighbouring blocks, and are idealized as a single or multiple finite elements. Block contacts are represented by a contact element. The DFEM utilizes the updated Lagrangian scheme so that it is possible large deformation of analysed domain resulting from rigid-body like motions of blocks. The equations of motion employing the principle of virtual work and conventional finite element discretization procedures are obtained for a typical finite element, in a condensed form, as follows:

$$M\ddot{U} + CU + KU = F \tag{3.38}$$

where,

$$M = \int_{\Omega e} \rho N^T N d\Omega; \quad C = \int_{\Omega e} B^T D_v B d\Omega; \quad K = \int_{\Omega e} B^T D_v B d\Omega; \quad F = \int_{\Omega e} N^T b d\Omega + \int_{\Gamma e} N^T t d\Gamma \tag{3.39}$$

This equation can be solved using one of the well-known techniques utilized in numerical analysis. If time domain is discretized, the final form of equation for a given time step takes the following form:

$$[\overline{K}]\{U\}_{n+1} = \{F\}_{n+1} \tag{3.40}$$

If central difference technique is adopted, the specific form of the equation above becomes:

$$[\overline{K}] = \frac{1}{2\Delta t}[M] + \frac{1}{2\Delta t}[C] \tag{3.41a}$$

$$\{F\}_{n+1} = \left(\frac{2}{\Delta t^2}[M] - [K]\right)\{U\}_n - \left(\frac{2}{\Delta t^2}[M] - \frac{2}{2\Delta t}[K]\right)\{U\}_{n-1} + \{F\}_n \tag{3.41b}$$

Assuming that contacts between two adjacent rock blocks has a certain thickness related to their roughness, it is modeled as shown in Figure 3.35 and its strain, strain rate are defined as follow:

$$\sigma_n = \frac{F_n}{A}, \quad \varepsilon_n = \frac{\delta_n}{h}, \quad \dot{\varepsilon}_n = \frac{\dot{\delta}_n}{h} \tau_s = \frac{F_s}{A}, \quad \gamma_s = \frac{\delta_s}{h}, \quad \dot{\gamma}_s = \frac{\dot{\delta}_s}{h} \tag{3.42}$$

Its finite element representation is illustrated in Figure 3.36.

There are three different approaches to deal with non-linear behaviour in numerical analyses, namely, initial stiffness, secant method and tangential stiffness method. The DFEM utilizes the initial stiffness or the secant technique (Figure 3.37). While the secant method can be much closer to the actual response, the ill-conditioning of global stiffness matrix may occur in very early steps of computation, the initial

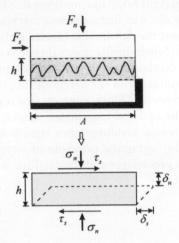

Figure 3.35 Mechanical modeling of contacts.

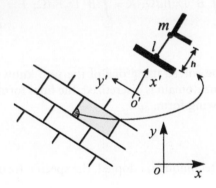

Figure 3.36 Finite element representation of a contact element.

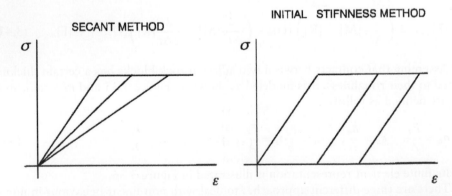

Figure 3.37 An illustration of techniques to deal with non-linearity in numerical analyses.

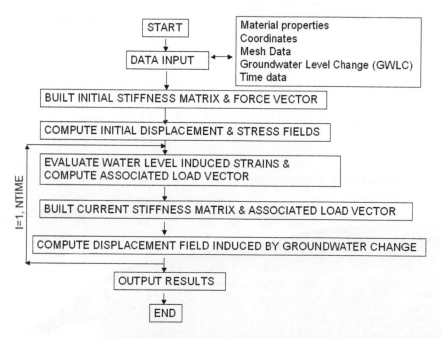

Figure 3.38 The flowchart of DFEM-CSH.

stiffness method yields much more numerically stable outputs. There are different versions of the DFEM, namely, hyperbolic, parabolic and pseudo-elliptic versions. While the original version is programmed as of hyperbolic type, the pseudo-elliptic type is more commonly used, as it is very difficult to have sufficient information on the elasto-visco-plastic behaviour of contacts, particularly.

As discussed in the previous subsection, it is impossible to model creep-like ground deformations at Gündoğdu district of Babadağ by classical sliding type models as the failure process would take place in several seconds. On the other hand, ground deformations have been taking place for decades in the district since the 1940s.

The fundamental concept described in Sub-Section 3.6.3.1 has been implemented in the discrete finite element method (DFEM). This version includes cyclic softening and hardening of contact zone and it is named as DFEM-CSH. The flowchart of the implementation of the DFEM code for this particular situation is shown in Figure 3.38.

Figures 3.39 and 3.40 show the principal stresses, maximum shear stress contours and deformed configurations for selected time steps for rainfall data starting from May 2011 continuing into 2012. As yielding of the contact zones was not allowed, the stress state remains the same while deformation of the body takes place upon each cycle of saturation and drying. As noted from Figure 3.40, very large displacement of the unstable zone does occur.

Figure 3.41 shows the response of three points at the rear, toe and middle-top of the potentially unstable body. As noted from the figure, the displacements at each

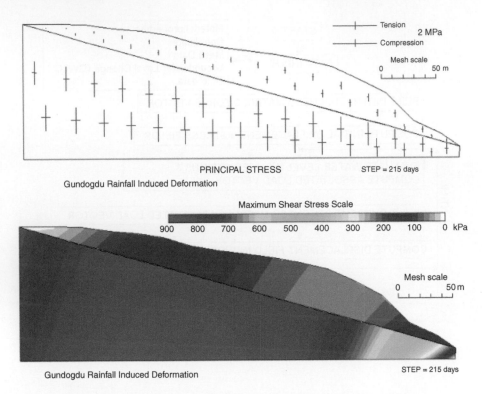

Gundogdu Rainfall Induced Deformation

Figure 3.39 Computed principal stress and maximum shear stress contours.

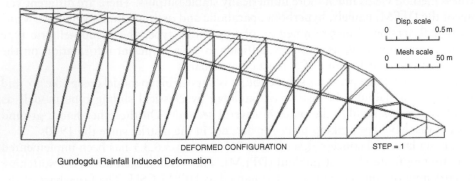

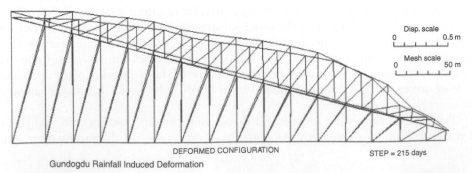

Figure 3.40 Deformed configurations of the analyzed domain at time steps 1 & 215 days.

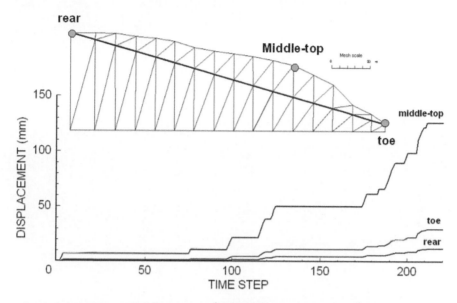

Figure 3.41 Displacement responses of three selected points in the analyzed domain.

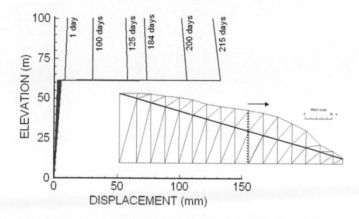

Figure 3.42 Shows ground deformation at a given section for different time-steps.

point differ and it is not purely a rigid-body-like ground deformation. The maximum ground deformation occurs at the middle top of about 150 mm for about 215 days.

Figure 3.42 shows horizontal ground deformation at a given section for different time-steps. It is very interesting to notice that the overall ground deformation resembles to those measured from pipe-strain gauge in the field shown in Figure 3.33.

Chapter 4

Thermo-mechanical behaviour of rocks and heat transport in rocks

4.1 INTRODUCTION

The heat flow through rock mass is of great importance in the utilisation of geothermal energy as well as in the assessment of the thermal environment in radioactive waste disposal. Thermal properties such as specific heat, heating or cooling coefficient and thermal conductivity are important to assess the heat transport through solids. In the first part of this chapter, the finite element presentation of the governing equations of heat transport derived is presented. Then, the theory for a simple testing method for thermal properties of solids is presented. In the final part, some applications of the finite element method to the modelling of geothermal state of a specific hot spring and faults are presented and compared with the actual measurements.

4.2 MECHANICAL MODELING HEAT TRANSPORT IN ROCKS

The well known governing equation of energy conservation in porous media takes the following form:

$$\rho c \frac{\partial T}{\partial t} = -\nabla \cdot (k \nabla T) + \boldsymbol{\sigma} \cdot \dot{\boldsymbol{\varepsilon}} + Q_h \tag{4.1}$$

The incremental form of the equation of motion is given by Eq. (4.2),

$$\nabla \cdot \dot{\boldsymbol{\sigma}} = 0 \tag{4.2}$$

The well-known equation of strain component induced by temperature variation is given by the following equation:

$$\dot{\boldsymbol{\varepsilon}}_T = \lambda \Delta T I \tag{4.3}$$

where I is the Kronecker Delta tensor. The constitutive law in terms of net-strain is generally written in the following incremental form:

$$\dot{\boldsymbol{\sigma}} = D(\dot{\boldsymbol{\varepsilon}} - \dot{\boldsymbol{\varepsilon}}_T) \tag{4.4}$$

4.3 NUMERICAL MODELING OF THERMO-MECHANICAL RESPONSES OF ROCKS

If one follows the conventional form of formulation based on the finite element method, governing equations (4.1) and (4.2) can be obtained as shown in the next subsections:

4.3.1 Weak form formulation

By taking a volumetric integration of Eq. (4.1) together with a variation δT on the temperature field and no heat production source, the weak form of Eq. (4.1) is easily obtained as

$$\int_V \rho c \delta T \frac{\partial T}{\partial t} dV + \int_V \nabla \delta T \cdot k \nabla T dV = \int_{S_q} \delta T \hat{h} dS + \int_V \delta T \cdot \dot{E} dV \tag{4.5}$$

where $\dot{E} = \boldsymbol{\sigma} \cdot \dot{\boldsymbol{\varepsilon}}$. The above equation is subjected to the following boundary conditions
Temperature boundary

$$T = T_0 \quad \text{on } S_T \tag{4.6}$$

Heat flux boundary

$$-(k \nabla T) \cdot \mathbf{n} = \hat{h} \quad \text{on } S_h \tag{4.7}$$

Since Eq. (4.5) holds for whole domain, it must also hold for sub-domains such as finite elements. Let us assume that the temperature field of a typical finite element can be interpolated through the following relation

$$T = [N]\{\chi\} \tag{4.8}$$

where $[N]$ is shape function. The dot product defined in mathematics is represented in the finite element method as follows:

$$\boldsymbol{\alpha} \cdot \boldsymbol{\beta} = \delta \quad \Rightarrow \quad \{\alpha\}^T \{\beta\} = \delta \tag{4.9}$$

With the use of above relations, the finite element form of Eq. (4.5) becomes

$$[M]_e \{\dot{\chi}\}_e + [K]_e \{\chi\}_e = \{Y\}_e \tag{4.10}$$

where

$$[M]_e = \int_V \rho c [N]^T [N] dV; \quad [K]_e = \int_V \lambda_T [B]^T [B] dV;$$

$$\{Y\}_e = \int_S [\overline{N}]^T \hat{h} dS + \int_V [N]^T \dot{E} dV$$

$$[M]\{\dot{X}\} + [K]\{X\} = \{Y\} \tag{4.11}$$

4.3.2 Discretization in time domain

Since Eq. (4.10) holds for any time, one may write the following for time step $(m + \theta)$ as

$$[M]\{\dot{X}\}_{(m+\theta)} + [K]\{X\}_{(m+\theta)} = \{Y\}_{(m+\theta)} \tag{4.12}$$

The Taylor expansion of variable $\{X\}$ for time step $(m + \theta)$ may be easily written as

$$\{X\}_m = \{X\}_{(m+\theta)-\theta} = \{X\}_{(m+\theta)} - \frac{\partial\{X\}_{(m+\theta)}}{\partial t}\frac{\theta\Delta t}{1!} + 0^2 \tag{4.13}$$

$$\{X\}_{m+1} = \{X\}_{(m+\theta)+(1-\theta)} = \{X\}_{(m+\theta)} + \frac{\partial\{X\}_{(m+\theta)}}{\partial t}\frac{(1-\theta)\Delta t}{1!} + 0^2 \tag{4.14}$$

Multiplying Eq. (4.13) by $(1 - \theta)$ and Eq. (4.14) by θ and summing up the resulting equations yield

$$\{X\}_{(m+\theta)} = \theta\{X\}_{m+1} + (1 - \theta)\{X\}_m \tag{4.15}$$

Furthermore, subtracting Eq. (4.14) from Eq. (4.15) results in

$$\{\dot{X}\}_{(m+\theta)} = \frac{\{X\}_{(m+1)} - \{X\}_m}{\Delta t} \tag{4.16}$$

Similarly $\{Y\}_{(m+\theta)}$ is obtained as

$$\{Y\}_{(m+\theta)} = \theta\{Y\}_{m+1} + (1 - \theta)\{Y\}_m \tag{4.17}$$

With the relations above, Eq. (4.12) may be re-written as

$$[C^*]\{X\}_{m+1} = \{Y^*\}_{m+1} \tag{4.18}$$

where

$$[C^*] = \left[\frac{1}{\Delta t}[M] + \theta[K]\right] \tag{4.19}$$

$$\{Y^*\}_{m+1} = \left[\frac{1}{\Delta t}[M] - (1 - \theta)[K]\right]\{X\}_m + \theta\{Y\}_{m+1} + (1 - \theta)\{Y\}_m \tag{4.20}$$

Summing up the above equations over the whole domain results in the following

$$[M]\{\dot{T}\} + [K]\{T\} = \{R\} \tag{4.21}$$

$$[K]\{\dot{U}\} = \{\dot{F}\} \tag{4.22}$$

where

$$[M] = S \int_V [N]^T [N] dV; \quad [K] = T \int_V [B]^T [B] dV; \quad [B] = \nabla [N];$$

$$\{R\} = \int_\Gamma [\overline{N}] q_n d\Gamma + \int_V [N] Q dV$$

$$[K] = \int_V [B]^T [D][B] dV; \quad \{\dot{F}\} = \int_V [B]^T [D]\{\dot{\varepsilon}_T\} dV$$

4.4   THERMAL PROPERTIES OF ROCKS AND THEIR MEASUREMENTS

Thermal properties such as specific heat, heating or cooling coefficient and thermal conductivity are important to assess the heat transport through solids as noted from Eq. (4.1). There are many techniques to measure thermal properties such as specific heat coefficient, thermal conductivity, thermal diffusion and thermal expansion coefficient. The details of such techniques can be found in various publications and textbooks (i.e. Clark, 1966; Somerton, 1992; Jumikis, 1993; Popov et al., 2016). Specific heat coefficient is commonly measured using the calorimeter tests.

The earlier and common technique for thermal conductivity measurement is the divided bar technique (Birch, 1950) based on the steady-state heat flow assumption and it is illustrated in Figure 4.1. This technique utilizes reference materials with well-known thermal properties.

There are also techniques for measuring thermal conductivity utilizing transient heat flow (i.e. Carslaw & Jaeger, 1959; Popov et al., 1999, 2016; Sass et al., 1984).

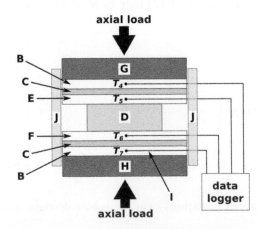

Figure 4.1   The key components of a divided-bar apparatus (from Popov et al., 2016).
   a – pivot point, b – brass disks, c – reference material, d – rock specimen, e – hot plate, f – cold plate, g – heat source (concealed Peltier device), h – heat sink, i – holes for the insertion of temperature sensors, j – thermal insulation.

These techniques utilize line or plane sources and temperature variations are measured by either contact sensor or infrared camera. These techniques utilize the analytical solutions developed by Carslaw & Jaeger (1959).

An experimental technique using a device similar a calorimeter type apparatus is described in this section in order to measure thermal properties of rock materials from a single experiment and its applications are given. Let us consider a solid (s) is enveloped by fluid (i.e. water (w)) as illustrated in Figure 4.2. It is assumed that solid and fluid have different thermal properties and temperature.

4.4.1 Definition of fundamental parameters

For theoretical modeling, the following parameters are defined as follow:

Q: heat; ρ: density; k: thermal conductivity: c: specific heat coefficient; T: temperature; m: mass; h: cooling coefficient; V: volume; A_s: surface area of solid; λ: thermal expansion coefficient.

The heat of a body is given in the following form

$$Q = m \cdot c \cdot T = \rho V \cdot c \cdot T \tag{4.23}$$

Its unit is Joule $(J = N \cdot m)$.

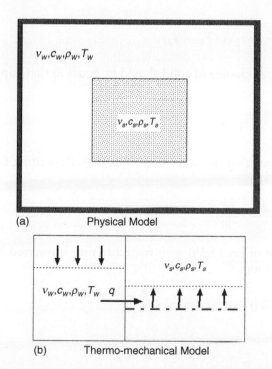

(a)   Physical Model

(b)   Thermo-mechanical Model

Figure 4.2 An illustration of physical and thermo-mechanical model.

Assuming that mass and specific heat coefficient are constant, the heat rate (heat flux) is given in the following form

$$\frac{dQ}{dt} = q = m \cdot c \cdot \frac{dT}{dt} \qquad (4.24)$$

Newton cooling law is written in the following form

$$q = h \cdot A_s \cdot \Delta T \qquad (4.25)$$

where ΔT: temperature difference between solid and enveloping fluid and its unit is Watt ($W = J/s$).

4.4.2 Physical model of experimental set-up

In this particular model, the temperature of surrounding fluid is assumed to be higher than the solid enveloped by the fluid. Furthermore, there is no heat flow from system outward. In other words it is thermally isolated. The heat flux from fluid can be given as

$$q_w = -\rho_w \cdot c_w \cdot V_w \frac{dT_w}{dt} \qquad (4.26)$$

The heat from fluid into solid should be equal with the use of Newton Cooling law as written below

$$-\rho_w \cdot c_w V_w \frac{\partial T_w}{\partial t} = h \cdot A_s(T_w - T_s) \qquad (4.27)$$

Similarly, the heat change of solid should be equal to that supplied from fluid as given by

$$\rho_s \cdot c_s \cdot V_s \frac{\partial T_s}{\partial t} = h \cdot A_s(T_w - T_s) \qquad (4.28)$$

It should be noted that the sign of heat flux is +. Re-writing Eq. (4.27) yields the following:

$$T_s = T_w + \frac{\rho_w c_w v_w}{h \cdot A_S} \cdot \frac{\partial T_w}{\partial t} \qquad (4.29)$$

If the derivation of Eq. (4.29) with respect to time is inserted into Eq. (4.25), one easily gets the following

$$\frac{\partial^2 T_w}{\partial t^2} + \alpha \frac{\partial T_w}{\partial t} = 0 \qquad (4.30)$$

where

$$\alpha = h \cdot A_s \frac{\rho_w \cdot c_w \cdot V_w + \rho_s \cdot c_s \cdot V_s}{\rho_w \cdot c_w \cdot V_w \cdot \rho_s \cdot c_s \cdot V_s} \qquad (4.31)$$

The solution of Eq. (4.30) would be obtained as follows:

$$T_w = C_1 + C_2 e^{-\alpha t} \tag{4.32}$$

Integral coefficients C_1 and C_2 of Eq. (4.32) are obtained from the following conditions:

$$T_w = T_i \quad \text{at } t = 0 \qquad \text{and} \qquad T_1 = T_f \quad \text{at } t = \infty \tag{4.33}$$

as

$$C_1 = T_f; \qquad C_2 = T_i - T_f \tag{4.34}$$

Using Integral constants given by Eq. (4.34), Eq. (4.32) becomes

$$T_w = T_f + \left(T_i - T_f\right) e^{-\alpha t}; \qquad \frac{\partial T_w}{\partial t} = -\alpha(T_i - T_f)e^{-\alpha t} \tag{4.35}$$

The average temperature of solid is obtained by inserting Eq. (4.36) into Eq. (4.30) as

$$T_s = T_f - \frac{\rho_w \cdot c_w \cdot V_w}{\rho_s \cdot C_s \cdot V_s}(T_i - T_f)e^{-\alpha t} \tag{4.36}$$

As $T_s = T_o$ at time $t = 0$, Eq. (4.36) can be re-written as

$$\frac{T_f - T_o}{T_i - T_f} = \frac{\rho_w \cdot c_w \cdot V_w}{\rho_s \cdot c_s \cdot V_s} \tag{4.37}$$

Inserting Eq. (4.37) into Eq. (4.35) yields

$$T_s = T_f - \left(T_f - T_o\right) e^{-\alpha t} \tag{4.38}$$

The temperature difference between solid and enveloping fluid can be obtained from Eqs. (4.35) and (4.38) as

$$\Delta T_{ws} = (T_w - T_s) = (T_i - T_o) e^{-\alpha t} \tag{4.39}$$

Therefore, if the values of T_o, T_f, T_i, ρ_w, ρ_s, V_w, V_s, c_w are known, the specific heat coefficient of solid can be easily obtained. For example, the specific heat coefficient of water is 4.1783–4.2174 J/g/K for a temperature range of 0–90°C. As the thermal properties of water remain almost constant for the given temperature range, the water would be used as fluid in the experimental set-up. After obtaining the specific heat coefficient, the coefficient α is obtained from Eq. (4.35), (4.38) or (4.39) using the curve fitting technique to experimental response. Then using the value of coefficient α, the value of cooling coefficient is obtained from Eq. (4.31).

For determining the thermal conductivity coefficient (k), the following approach is used. Fourier law may be written for one-dimensional situation can be written as

$$q = -kA\frac{\partial T}{\partial x} \tag{4.40}$$

Assuming the specimen has a length (L) and using the Newton's cooling law we may write the following relationship:

$$hA\Delta T = kA\frac{\Delta T}{L} \tag{4.41}$$

Eq. (4.41) can be re-written and the following relation holds between cooling coefficient and thermal conductivity

$$k = h \cdot L \tag{4.42}$$

The characteristics length of a solid sample can be obtained from the volume of the solid from the following relationship:

$$L = \sqrt[3]{V_s} \tag{4.43}$$

Linear thermal expansion coefficient (λ) is defined as:

$$\lambda = \frac{1}{L} \cdot \frac{dL}{dT} \tag{4.44}$$

where L: length of sample. $\frac{dL}{dT}$ is the variation of length of sample with respect to temperature variation and It is determined under the unstrained condition or $100\,\mathrm{gf}$ load on the sample. If the variation of length of sample at the equilibrium state with respect to the initial length before the commencement of the experiment is measured, it would be straightforward to obtain linear expansion coefficient. Similarly width or diametrical changes can be also measured and thermal expansion coefficients can be evaluated from the variation of side length or diameter for given temperature difference.

4.4.3 Experimental procedure

The technique described in this sub-section is unique and quite practical considering the labourship in other techniques.

The device for determining the thermal properties of geo-materials consists of a thermostat cell equipped with temperature sensors. The fundamental features of this device are illustrated in Figure 4.3 and its picture is given in Figure 4.4. In the experiments, the temperature of sample, water, air and thermostat are measured. The method utilizes the thermal properties of water, whose properties remain to be the same up to 90 °C, to infer the thermal properties of geo-material substances. If the continuous

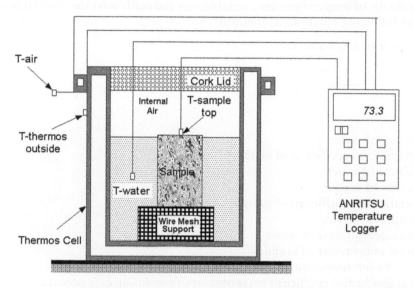

Figure 4.3 Illustration of experimental set-up.

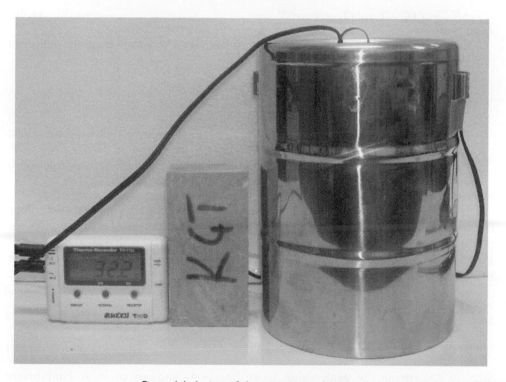

Figure 4.4 A view of the experimental set-up.

measurements of temperatures are available, one can easily infer the thermal properties from the following equations as follows:

Specific heat of geo-material

$$c_s = \frac{\rho_w \cdot c_w \cdot V_w}{\rho_s \cdot V_s} \frac{T_i - T_f}{T_f - T_o} \tag{4.45}$$

where
ρ_w : density of water
c_w : specific heat coefficient of water
V_w : Volume of water
ρ_s : density of sample
c_s : specific heat coefficient of sample
V_s : Volume of sample
T_i : initial temperature of water
T_o : initial temperature of sample
T_f : equilibrium temperature
    Heat conduction coefficient (α) is obtained from fitting experimental results to the following equation

$$\Delta T_{ws} = (T_w - T_s) = (T_i - T_o) \, e^{-\alpha t} \tag{4.46}$$

If heat conduction coefficient (α) is determined, then Newton's cooling coefficient is determined from the following equation

$$h = \frac{\alpha}{A_s} \cdot \frac{\rho_w \cdot c_w \cdot V_w \cdot \rho_s \cdot c_s \cdot V_s}{\rho_w \cdot c_w \cdot V_w + \rho_s \cdot c_s \cdot V_s} \tag{4.47}$$

Finally, thermal conductivity coefficient is obtained from the following equation

$$k = h \cdot L \tag{4.48}$$

where L is the characteristics sample side length
    If the sample temperature can be measured, it will be very easy to determine the specific heat coefficient of sample and subsequent properties. This is possible for granular materials since the temperature sensor can be embedded in the center of samples. However, it is quite difficult to determine the equilibrium temperature T_f for solid samples. Therefore, the following procedure is followed for this purpose:

Step 1: Determine the heat conduction coefficient (α)
Step 2: Plot the following equation in time space

$$T_s = T_w - (T_i - T_o)e^{-\alpha t} \tag{4.49}$$

Step 3: Determine the peak value from Eq. (4.4) and assign it as equilibrium temperature T_f.
Step 4: Then proceed to determine the rest of thermal properties using the procedure described above.

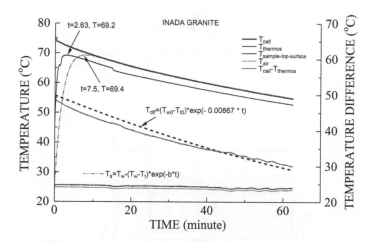

Figure 4.5 Application of the procedure to Inada granite sample.

Figure 4.5 shows the application of the method to a cylindrical Inada granite sample. The temperature of the sample at the top was also measured. As noted from the figure, the sample temperature at the top of sample achieves the peak value before the computed response. Since the temperature of the sample averaged over the total volume of the sample in the proposed temperature, the computed sample temperature achieves its peak value later than that at the top surface of the sample.

4.5 APPLICATIONS

4.5.1 Temperature evolution in rock due to hydration of adjacent concrete lining

Concrete linings of various thicknesses are usually constructed for acquiring dry working conditions in structures such as shafts, tunnels etc. throughout their service life as well as for their stability. If such structures are excavated through water bearing strata, the permeability of the concrete linings becomes very important as it governs the water inflow into the openings. Although concrete itself may be regarded as an impermeable material practically from the engineering point of view, the water inflow through linings would be mostly due to cracks which may exist in those. The cracks which are often observed on sites may result from various causes such as ground and/or water pressures and thermal stresses developing during the hydration process, etc. As these cracks will cause various undesirable problems such as water pumping particularly in deep shafts and/or tunnels below the ground water table and even instability, counter measures against the occurrence of those should be undertaken.

In this section, two specific examples on the evolution of temperature distributions with time in concrete linings of shafts and tunnels and the surrounding medium are described. The first example is associated with the temperature evolution in the concrete lining of Muna Tunnels in Saudi Arabia during its hydration process. A physical model set-up of the concrete lining and surrounding rock was prepared in the

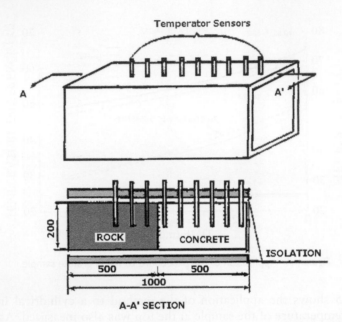

Figure 4.6 A physical model set-up of the concrete lining and surrounding rock of Muna Tunnels.

Table 4.1 Thermo-physical properties of materials used in simulations.

Material	k (kcal/m hr °C)	c (kcal/kg °C)	ρ (kg/m³)	α (m²/hr)
Shutter	0.1022	0.668	510	0.0003
Concrete	3.187	0.22	2300	0.0062
Rock	2.28	0.146	2500	0.0062

laboratory as shown in Figure 4.6 (Aydan *et al.*, 1986). The model was 1000 mm long and 200 mm high and 200 mm wide. The surrounding rock was granite sampled from the Muna tunnels and 9 temperature sensors were embedded in concrete and rock as shown in Figure 4.6. The heat production of the concrete mix was measured and used as input data. Table 4.1 gives thermo-physical properties of materials of the physical model used in numerical simulations. Figure 4.7 compares the computed temperature distributions in concrete and surrounding rock at chosen time intervals. The maximum temperature rise occurs 12 hours after the casting of the concrete and temperature of concrete and surrounding rock tend to decrease as time goes by.

The next example is concerned with the temperature distribution of concrete linings of the 1000 m deep shafts of North Selby Mine, Yorkshire, UK (Aydan, 1989). The shafts were to be excavated through a thick water-bearing sandstone formation. As the sandstone formation was a major source in the area, it was decided to adopt the ground freezing method to prevent water inflow into the shafts during construction.

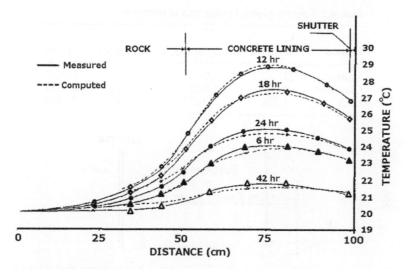

Figure 4.7 Comparison of measured and computed temperature distributions during the hydration process of concrete.

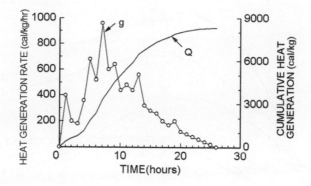

Figure 4.8 Heat generation function of the concrete mix used in North Selby Mine Shaft.

The frozen collar of rock extends up to 280 m below the ground surface at the deepest location. When thick concrete linings cast against frozen ground it is customary to supplement an additional 150 mm thickness to the lining to account into the incomplete hydration process of concrete. As the shafts were very deep, the thickness of concrete linings were increased. At certain locations, this thickness was more than 1300 mm. An experimental program was undertaken to measure temperature variations, strains in linings and water and ground pressures during and after construction. The objectives were to check the hydration process of concrete linings cast against frozen sandstone and thermal cracking problem in thick concrete linings cast against frozen or unfrozen rock masses.

A series of experiments on heat generation rate of cement, which was a rapid-hardening type, and thermo-mechanical properties of concrete and rock were undertaken. Figure 4.8 shows the heat generation function of the concrete mix. Table 4.2

Table 4.2 Thermo-physical properties of materials.

Material	k (kcal/m hr °C)	c (kcal/kg °C)	ρ (kg/m³)	α (m²/hr)
Shutter	32	0.153	7800	0.0267
Concrete	2.088	0.22	2300	0.0033
Rock	1.728	0.257	2171	0.0025

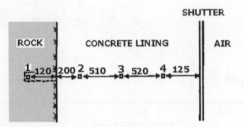

Figure 4.9 Instrumentation of the concrete lining and surrounding rock the shaft of the North Selby Mine.

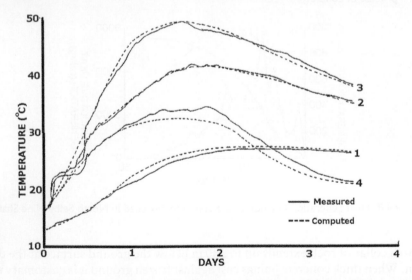

Figure 4.10 Comparison of measured and computed temperature distributions in concrete lining during its hydration process.

gives thermo-mechanical properties of the shutter, concrete lining and rock mass. The inner diameter of a typical lined shaft was 7315 mm and the concrete was poured against the strata in 6 m lengths. Taking into account the fact, the problem is modelled as an axisymmetric problem in finite element analysis. A computation was carried out for concrete lining where its thickness was 1355 mm and it was instrumented (Figure 4.9). Figure 4.10 shows the calculated and measured results at gauges 1, 2, 3 and 4. The finite element analysis predicted a maximum temperature of 46° near the

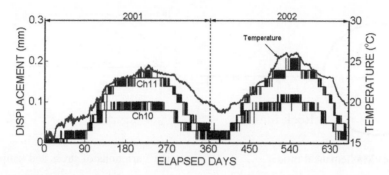

Figure 4.11 Displacement responses together with variation of temperature during the period between 2001 and 2002.

Table 4.3 Thermo-hydro-mechanical properties of surrounding rock mass.

Unit weight (kN/m^3)	Elastic Modulus (GPa)	Poisson's ratio	Thermal Diffusivity (m^2/day)	Thermal Expansion Coefficient $(1/°C)$
26	5–10	0.25	0.1	1.0×10^{-5}

middle of the lining at 30 hrs following the concrete placement while the actual one was order of 50° and occurred 29.4 hrs after the placement. The difference may be attributed to the local variation of the amount of cement and local non-uniformity of temperature in concrete mixture. Nevertheless, It may be said the both results are in good agreement with each other. However, the measured temperature curves of gauges 1, 2 and 3 are not so smooth as the calculated ones, particularly in the first 12 hours. This was considered to be due to sudden increases and decreases in heat generation rate of cement paste in the initial stages of hydration process. The experimental and calculated results furthermore indicated that the temperature distribution of the concrete lining remained above the freezing temperature for about 5 days which was a sufficient period of time for the completion of the hydration period of the concrete.

4.5.2 Underground cavern in rock

Aydan *et al.* (2012) reported the temperature and humidity measurements in the cavern and at the entrance of the powerhouse since 2006 (Figure 4.11). The temperatures are different in the cavern and the yearlong variation is within 8–10 degrees. The daily variation ranges between 2–3 degrees. However, the variation at the entrance is quite large and the yearlong variation is about 30–32 degrees while the daily variation is around 8–12 degrees. The flow of water through the turbines of the powerhouse also has some effect on the temperature field in the close vicinity of the turbines.

In the first series of the thermo-mechanical finite element analyses, the cavern was modelled as a one-dimensional axisymmetric cavity with the use of parameters given in Table 4.3 and equivalent area. The cavern was subjected to ±10° yearlong

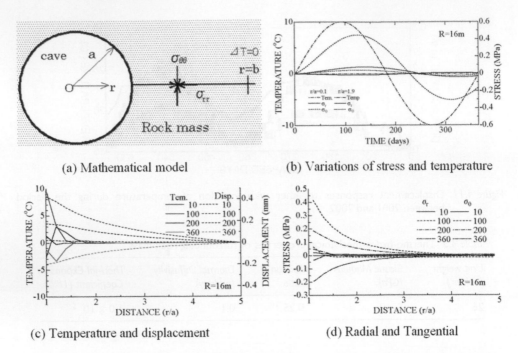

(a) Mathematical model

(b) Variations of stress and temperature

(c) Temperature and displacement

(d) Radial and Tangential

Figure 4.12 Computed results for one-dimensional axisymmetric thermo-mechanical finite element analysis.

sinusoidal temperature variation. This thermal environment is probably the extremist case of the temperature variation in the cavern. Figure 4.12 shows the computed results. The tangential stress variation in the close vicinity of the cavern is the highest and its value is about 0.4 MPa. This implies that the rock mass adjacent to the cavern wall will be subjected to ±0.4 MPa thermal stress cycles. The amplitude of these cycles disappears when the distance from the cavern wall becomes three times the cavern radius.

Two-dimensional finite element analyses were carried out with the application of the same temperature variation on the cavern surface (Figures 4.13, 4.14 and 4.15). Figure 4.13 shows responses of stresses and displacements for RHS and LHS walls and crown.

Figure 4.14 shows temperature distribution, principal stresses and maximum shear stress distributions. The results shown in these figures are basically similar to the one-dimensional axisymmetric finite element analyses in terms of amplitudes. Nevertheless, the effect of cavern geometry on stress and displacement responses is noted.

Figure 4.15 compares measured displacements with computed displacements at spring lines of cavern walls. As noted from the figure, the responses and amplitudes of computed displacements with time are quite similar to those of the measured displacements.

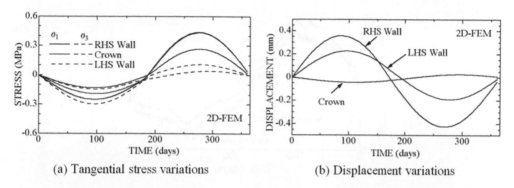

(a) Tangential stress variations   (b) Displacement variations

Figure 4.13 Stress and displacement variations at selected points around the cavern.

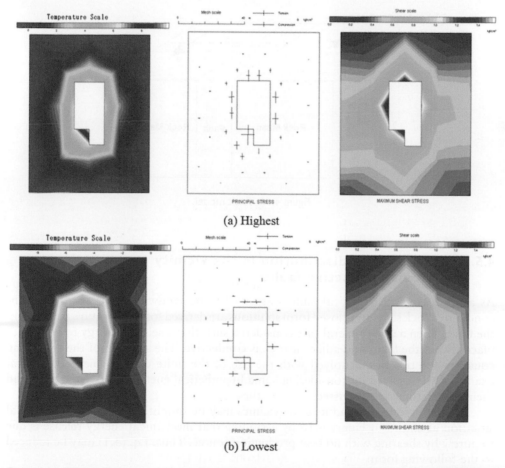

Figure 4.14 Temperature, principal stresses and maximum shear stress distributions (two-dimensional analyses).

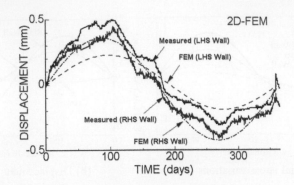

Figure 4.15 Comparison of computed results for two-dimensional thermo-mechanical finite element analysis with measured displacement at the spring-line of the cavern.

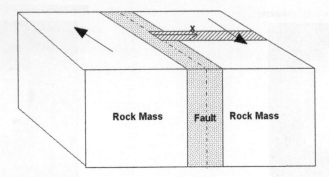

Figure 4.16 Fault model.

4.5.3 Temperature distribution in the vicinity of geological active faults

As a first case, the geological fault is assumed to be sandwiched between two non-conductive rock slabs and closed form solutions are derived for temperature rises within the fault. Then a more general case considered such that a seismic energy release takes place within the fault, and adjacent rock is conductive. The solution of the governing equation for this case is solved with the use of the finite element method. Several examples were solved by considering some hypothetical energy release functions and their implications are discussed.

If a geological fault and its close vicinity may be simplified to a one-dimensional situation as shown in Figure 4.16 by assuming that mechanical energy release is due to purely by shearing with no heat production source. Thus, Eq. (4.1) may be reduced to the following form:

$$\rho c \frac{\partial T}{\partial t} = -\nabla q + \tau \dot{\gamma} \tag{4.50}$$

Let us assume that the heat flux obeys to Fourier's law, which is given by

$$q = -k\frac{\partial T}{\partial x} \tag{4.51}$$

Inserting Eq. (4.51) into Eq. (4.50) yields the following equation

$$\rho c\frac{\partial T}{\partial t} = k\frac{\partial^2 T}{\partial x^2} + \tau\dot{\gamma} \tag{4.52}$$

The solution of the above equation will yield the temperature variation with time.

The energy release during earthquakes is very complex phenomenon. Nevertheless, some simple forms relevant for the overall behaviour may be assumed in order to have some insight to the phenomenon. Two energy release rate functions of the following form are assumed as given below:

$$\dot{E} = \tau\dot{\gamma} = Ate^{-\frac{t}{\theta}} \tag{4.53}$$

$$\dot{E} = \tau\dot{\gamma} = A^* e^{-\frac{t}{\theta^*}} \tag{4.54}$$

Constants A and A^* depend on the shear stress and shear strain rate history with time and fault thickness. Constants θ and θ^* are time history constants. For a situation illustrated in Figure 4.15, constants A and A^* will take the following forms:

For Eq. (4.53) $A = \dfrac{\tau_o u_f}{h\theta^2}$ \hfill (4.54)

For Eq. (4.54) $A^* = \dfrac{\tau_o u_f}{h\theta^*}$ \hfill (4.55)

where u_f, h are final relative displacement and thickness of the fault. τ_o is the shear stress acting on the fault and it is assumed to be constant during the motion.

Two specific situations are analysed, namely:

- Creeping Fault
- Fault with hill-shaped seismic energy release rate

In the case of creeping fault, the energy release rate is almost constant with time. The geometry of the fault is assumed to be one-dimensional as shown in Figure 4.17. Figures 4.18 and 4.19 show the computed temperature differences at selected locations with time and temperature difference distribution throughout the whole domain at selected time steps. In the computations, the energy release rate is assumed to be taking place within the fault zone only. The increase of temperature difference is parabolic and they keep increasing as time goes by. Nevertheless, the temperature difference increases are about 1/10 of those of the fault sandwiched between non-conductive rock mass slabs.

Figures 4.20 and 4.21 show the computed temperature differences at selected locations with time and temperature difference distribution throughout the whole domain

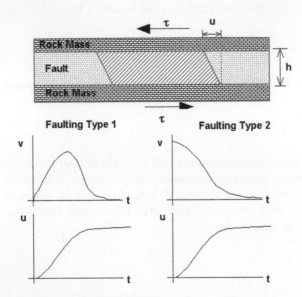

Figure 4.17 Faulting models and energy release types.

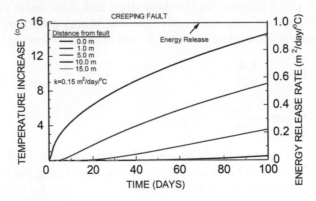

Figure 4.18 Temperature difference variations for a fault sandwiched between conductive rock mass slabs for creeping condition.

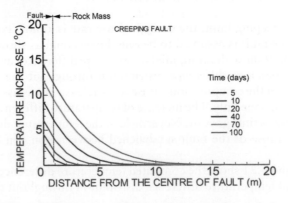

Figure 4.19 Temperature distributions at different time steps for a fault sandwiched between conductive rock mass slabs for creeping condition.

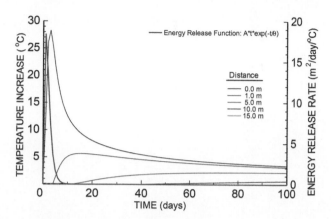

Figure 4.20 Temperature difference variations for a fault sandwiched between conductive rock mass slabs for hill-shaped energy release function.

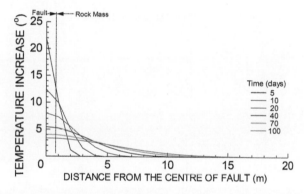

Figure 4.21 Temperature distributions at different time steps for a fault sandwiched between conductive rock mass slabs for hill-shaped energy release function.

at selected time steps for a fault with a hill-like energy release rate. In the computations, the energy release rate is assumed to be taking place within the fault zone only. The increase of temperature difference is parabolic. Temperature difference increases first and then they tend to decay in a similar manner to the assumed seismic energy release rate function. This situation will be probably quite similar to the actual situation in nature. The temperature difference increases are about 1/10 of those of the fault sandwiched between non-conductive rock mass slabs. These results indicate that the observation of ground temperatures may be very valuable source of information in the predictions of earthquakes. Because of atmospheric temperature measurements near the ground surface may be quite problematic in interpreting the observations. However, the observation of hot-spring temperature, which reflects the actual ground temperature, may be very good tool for such measurements without any deep boring.

Chapter 5

Hydromechanics of rocks and rock engineering structures

5.1 INTRODUCTION

The effect of water is of great importance for assessing the mechanical response of geotechnical engineering structures during and after construction. It is well known that groundwater causes various stability problems for rock engineering structures. Some of well known examples are shown in Figure 5.1. The overtopping of the reservoir water in the Vaiont dam was caused by a landslide and resulted in heavy casualties down-stream. The failure of the foundation of the Malpasset dam also caused severe

Figure 5.1 Examples of seepage through rock mass and some associated problems.

casualties on the downstream side of the dam. The excavation of tunnels causes the ground subsidence besides local stabilities within the tunnels. The ground above the Nakayama Tunnel in Japan subsided more than 1.5 m during the reduction of ground water level to deal with heavy water-inflow into the tunnel.

It is well known that ground water adversely affects the stability of excavations as well as the working conditions (Aydan & Ersen, 1983). In recent years, studies on the disposal of radioactive wastes in rockmass have been increasing in the countries where a great proportion of energy production is based on nuclear power plants. In case of any accidence during disposing these wastes in rock mass, the transportation of the radioactive wastes to the ecological environment strictly depends upon the seepage characteristics of rock mass (Brace *et al.*, 1968; Gale, 1990; Tsang & Witherspoon, 1981). Furthermore, the productivity of wells in petroleum industry also depends upon the seepage characteristics of rock mass.

To consider the mechanical effect of water on soils, Terzaghi (1925) introduced the concept of the effective stress. This concept was later also used for rocks. On the other hand, Biot (1942) introduced a general effective stress law which incorporates the volumetric porosity n and the ratio of the volumetric stiffness of solid K_s and that of the bulk K. Although this law is a more general one and it is the basis of the mixture theory for porous media, Terzaghi's effective law is widely accepted in geotechnical engineering as it is very simple and it does not involve any deformability characteristics of the solid and fluid phases.

Rock mass in nature is generally fractured. As a result, they contain numerous discontinuities, which may be very important in the stability of rock engineering structures and mass transport through the rock mass. The same effective stress laws have been also extended to the rock discontinuities (Byerlee, 1967). However, there are very few experimental studies to check the applicability and validity of the effective stress laws for rock discontinuities.

This chapter explain fundamental aspects on the hydromechanics of rocks and how to solve the fundamentals of governing equations and present several applications in practice.

5.2 FUNDAMENTAL EQUATION OF FLUID FLOW IN POROUS MEDIA

The mass conservation law for fluid flowing through the pores within rock may be given in the following form with the use of the mixture theory and assuming that a coordinate system fixed to the solid phase (i.e. Aydan, 2000a,b):

$$\frac{\partial (\phi \rho_f)}{\partial t} = -\nabla \cdot (\phi \mathbf{q}_f) \tag{5.1}$$

where $\nabla = \frac{\partial}{\partial x_i} \mathbf{e}_i$, $i = 1, 3$; ρ_f: fluid density, ϕ: porosity, $\mathbf{q}_f$: fluid flux.

One may write the following relation for fluid flux in terms of relative velocity $\mathbf{v}_r$ of fluid and the velocity $\mathbf{v}_s$ of solid phase as

$$\phi \mathbf{q}_f = \rho_f (\mathbf{v}_r + \phi \mathbf{v}_s) \tag{5.2}$$

Let us assume that the flow of fluid obeys the Darcy's law. Thus we have the following

$$\mathbf{v}_r = -\frac{k}{\eta}\nabla p \tag{5.3}$$

where k is permeability, η is viscosity of fluid. Inserting Eqs. (5.3) and (5.2) into Eq. (5.1) yields the following

$$\frac{\partial(\phi\rho_f)}{\partial t} = \nabla \cdot \left(\rho_f \left(\frac{k}{\eta}\nabla p - -\phi\mathbf{v}_s \right) \right) \tag{5.4}$$

The material derivative operator according to Eulerian description may be written as (Eringen, 1980):

$$\frac{d_s}{dt} = \frac{\partial}{\partial t} + \mathbf{v}_s \cdot \nabla(\ ) \tag{5.5}$$

Introducing this operator into Eq. (5.4), we have the following relation

$$\frac{d_s\phi}{dt} + \frac{\phi}{\rho_f}\frac{d_s\rho_f}{dt} = \nabla \cdot \left(\frac{k}{\eta}\nabla p \right) - \frac{1}{\rho_f}\nabla \cdot (\phi\mathbf{v}_s) \tag{5.6}$$

The following constitutive relations are assumed to hold among porosity, fluid and solid densities and pressure (i.e. Zimmerman et al., 1986)

$$\frac{d_s\phi}{dt} = (C_b - (1+\phi)C_s)\frac{d_s p}{dt}; \quad \frac{1}{\rho_f}\frac{d_s\rho_f}{dt} = C_f\frac{d_s p}{dt} \tag{5.7}$$

If the velocity of solid phase is assumed to be small so that it can be neglected, Eq. (5.6) takes the following form with the use of Eq. (5.7):

$$\beta\frac{\partial p}{\partial t} = \nabla \cdot (\nabla p) \tag{5.8}$$

where

$$\beta = [(C_b - C_s) + \phi(C_f - C_s)]\frac{\eta}{k} \tag{5.9}$$

5.2.1 Special form of governing equation

Eq. (5.8) can be re-written for one-dimensional longitudinal flow as:

$$\beta\frac{\partial p}{\partial t} = \frac{\partial^2 p}{\partial x^2} \tag{5.10}$$

Similarly Eq. (5.8) can be also written for axi-symmetric radial flow as:

$$\beta\frac{\partial p}{\partial t} = \frac{1}{r}\frac{\partial}{\partial r}\left(r\frac{\partial p}{\partial r} \right) \tag{5.11}$$

5.2.2   Governing equations of fluid in reservoirs attached to sample

Using the mass conservation law and the constitutive relation between pressure and fluid density, the velocities v_1, v_2 of fluid contained in reservoirs numbered (1) and (2) and attached to the ends of a sample can be written as:

$$v_1 = -C_f V_1 \frac{\partial p_1}{\partial t}, \quad v_2 = -C_f V_2 \frac{\partial p_2}{\partial t} \tag{5.12}$$

where V_1 and V_2 are volumes of reservoirs, and p_1 and p_2 are pressures acting on fluid reservoirs.

5.3   PERMEABILITY CHARACTERISTICS OF ROCKS AND DISCONTINUITIES

To measure the seepage characteristics of rock masses, numerous methods are proposed, and the preference to select an appropriate method is generally associated with the expected permeability values of rock masses (Aydan *et al.*, 1997b). When the expected permeability of rock mass is relatively large, constant head or falling head permeability tests, which are widely used in soil mechanics, are employed. On the other hand, the transient pulse test proposed by Brace *et al.* (1968) is used when the permeability of rock masses is relatively small. The permeability is obtained from time-pressure difference relation observed in this test. Aydan *et al.* (1997b) extended this method to axi-symmetric radial flow tests. The compressibility of solid and fluid phases of samples is not taken into account in interpreting the test results in this method. It is natural to expect that the permeability values may be different if the compressibility of the samples is taken into account. There are two kinds of laboratory tests available for measuring permeability of rock specimen, namely, steady-state and non-steady state flow tests.

5.3.1   Some considerations on Darcy law for rocks and discontinuities

Darcy law (given in Eq. (5.3)) is generally used as a constitutive model for the fluid flow through porous rock and rock discontinuities together with the assumption of laminar flow. A brief description of Darcy's law is presented in this sub-section.

Darcy performed a series of experiments on a sand column in 1856. From these experiments, he found out that the volume discharge rate Q is directly proportional to the head drop $h_2 - h_1$ and to the cross-sectional area A, but it is inversely proportional to the length difference $l_2 - l_1$. Calling the proportionality constant K as the hydraulic conductivity, Darcy's law is written:

$$Q = -KA \frac{h_2 - h_1}{l_2 - l_1} \tag{5.13}$$

The negative sign signifies that groundwater flows in the direction of head loss.

Darcy's law is now widely accepted and used in modelling fluid flow in porous or fractured media. It is elaborated and written in a differential form which is given below for one-dimensional case as:

$$v = -K\frac{\partial h}{\partial x} \tag{5.14}$$

This law is analogous to Fourier's law in heat flow presented in Chapter 4. Darcy's law is theoretically derived for tube-like pores and slit-like discontinuities in this subsection (Aydan *et al.*, 1997a; Aydan & Üçpırtı, 1997).

(a) Darcy law for rock with cylindrical pores
Equilibrium equation for x-direction is given as

$$\sum F_x = p\pi[(r + \Delta r)^2 - r^2] - (p + \Delta p)\pi[(r + \Delta r)^2 - r^2] \tag{5.15}$$
$$+ (\tau + \Delta\tau)2\pi(r + \Delta r)\Delta x - \tau 2\pi r\Delta x = 0$$

Re-arranging the resulting expression and taking the limit and omitting the second order components yields:

$$\frac{dp}{dx} - \frac{d\tau}{dr} - \frac{\tau}{r} = 0 \tag{5.16}$$

Assuming that the flow is laminar and a linear relationship holds between shear stress and strain rate $\dot{\gamma}$ as:

$$\tau = \eta\dot{\gamma}, \quad \dot{\gamma} = \frac{d\dot{u}}{dr} = \frac{dv}{dr}, \quad v = \dot{u} = \frac{du}{dt} \tag{5.17}$$

Now, let us insert the above relation into Eq. (5.16), we have the following partial differential equation:

$$\frac{dp}{dx} - \eta\frac{d^2v}{dr^2} - \frac{\eta}{r}\frac{dv}{dr} = 0 \tag{5.18}$$

Integrating the above partial differential equation for *r*-direction yields the following

$$v = \frac{1}{\eta}\frac{dp}{dx}\frac{r^2}{4} + C_1 \ln r + C_2 \tag{5.19}$$

Introducing the following boundary conditions as:

$$v = v_0 \quad \text{at } r = \frac{D}{2}$$
$$\tau = 0 \quad \text{at } r = 0$$

yields the integration constants C_1 and C_2 as:

$$C_1 = 0, \quad C_2 = v_0 - \frac{1}{\eta} \frac{dp}{dx} \frac{D^2}{16}$$

where D is diameter of pore. If velocity v_0 is given in the following form

$$v_0 = -\alpha \frac{1}{\eta} \frac{dp}{dx} \frac{D^2}{16} \tag{5.20}$$

integration coefficient C_2 can be obtained as follows

$$C_2 = -(1+\alpha) \frac{1}{\eta} \frac{dp}{dx} \frac{D^2}{16}$$

The flow rate q passing through the discontinuity for a unit time is:

$$q = \int_0^{2\pi} \int_{r=0}^{y=\frac{D}{2}} vr \, dr \, d\theta \tag{5.21}$$

The explicit form of q is obtained as

$$q = -\frac{\pi}{\eta} \frac{D^4}{128} \frac{dp}{dx} \tag{5.22}$$

If the flow rate q is re-defined in terms of an average velocity $\bar{v}$ over the pore area as

$$q = -\bar{v} \pi \frac{D^2}{4} \tag{5.23}$$

we have the following expression

$$\bar{v} = -(1+\alpha) \frac{1}{\eta} \frac{D^2}{32} \frac{dp}{dx} \tag{5.24}$$

This relation is known as Hagen-Poiseuille for $\alpha = 0$. In an analogy to the Darcy law, we can re-write the above expression as

$$\bar{v} = -\frac{k}{\eta} \frac{dp}{dx} \tag{5.25}$$

where

$$k = (1+\alpha) \frac{D^2}{32} \quad \text{or} \quad k = (1+\alpha) \frac{a^2}{8}; \quad a = \frac{D}{2}$$

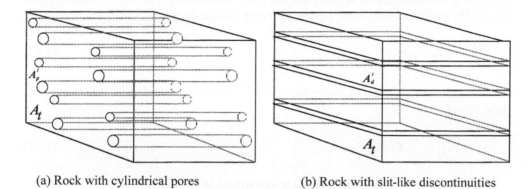

(a) Rock with cylindrical pores   (b) Rock with slit-like discontinuities

Figure 5.2 Geometrical models for Darcy law.

This is known as the actual permeability of the pores. Let us assume that the ratio (porosity) n of the area of pores over the total area is given by (Figure 5.2a)

$$n = \frac{1}{A_t} \sum_{i=1}^{N} \pi \frac{D_i^2}{4} \quad \text{or} \quad n = \frac{N\pi \overline{D}^2}{4A_t} \tag{5.26}$$

Then, the apparent permeability k_a is related to the actual permeability as

$$k_a = nk \tag{5.27}$$

(b) Darcy law for slit-like discontinuities

For x-direction, force equilibrium equation for fluid can be given as follows (Figure 5.2b):

$$\sum F_x = p_{(x)}\Delta y - p_{(x+\Delta x)}\Delta Y + \tau_{(y+\Delta y)}\Delta x - \tau_{(y)}\Delta x = 0 \tag{5.28}$$

where p is pressure and τ is shear stress. Eq. (5.28) takes the following partial differential form by taking Taylor expansions of p and τ as:

$$\frac{dp}{dx} - \frac{d\tau}{dy} = 0 \tag{5.29}$$

Assuming that flow is laminar and the relation between shear stress τ and shear strain rate $\dot{\gamma}$ is linear:

$$\tau = \eta\dot{\gamma}, \quad \dot{\gamma} = \frac{d\dot{u}}{dy} = \frac{dv}{dy}, \quad v = \dot{u} = \frac{du}{dt} \tag{5.30}$$

where η is viscosity and $\dot{u}$ is deformation rate. Substituting the above relations into Eq. (5.29) yields the following partial differential equation:

$$\frac{dp}{dx} - \eta\frac{d^2v}{dy^2} = 0 \tag{5.31}$$

Integrating the equation above for y-direction yields the following expression for flow velocity v

$$v = \frac{1}{\eta}\frac{dp}{dx}\frac{y^2}{2} + C_1 y + C_2 \tag{5.32}$$

Introducing the following boundary conditions in Eq. (5.32)

$$v = v_o \quad \text{at } y = \frac{h}{2}, \quad \tau = 0 \quad \text{at } y = 0$$

yields the integration constants C_1 and C_2 as:

$$C_1 = 0, \quad C_2 = v_o - \frac{1}{\eta}\frac{dp}{dx}\frac{h^2}{8} \tag{5.33}$$

where h is the aperture of discontinuity. If it is assumed that the following relation exists for v_o

$$v_o = -\alpha\frac{1}{\eta}\frac{dp}{dx}\frac{h^2}{8} \tag{5.34}$$

then, the integration constant C_2 can be written as:

$$C_2 = -(1+\alpha)\frac{1}{\eta}\frac{dp}{dx}\frac{h^2}{8} \tag{5.35}$$

Total flow rate v_t through the discontinuity at a given time is:

$$v_t = 2\int_{y=0}^{y=\frac{h}{2}} v\,dy \tag{5.36}$$

The explicit form of v_t is obtained as:

$$v_t = -(1+\alpha)\frac{1}{\eta}\frac{h^3}{12}\frac{dp}{dx} \tag{5.37}$$

For $\alpha = 0$, the above equation is well known as a "cubic law" equation in groundwater hydrology (Snow, 1968) and it is introduced to the field of geomechanics by Polubarinova-Kochina in 1962. Let us re-define the flow rate v_t in terms of an average velocity $\bar{v}$ and the discontinuity aperture h as

$$v_t = \bar{v}h \tag{5.38}$$

Inserting this equation into Eq. (2.26) yields the following:

$$\bar{v} = -\frac{1}{\eta}\frac{b^2}{12}\frac{dp}{dx}$$

(5.39)

In an analogy to the Darcy law, the above equation may be re-written as:

$$\bar{v} = -\frac{k_d}{\eta}\frac{dp}{dx}$$

(5.40)

where

$$k_d = (1+\alpha)\frac{b^2}{12}$$

k_d in the above equation is called permeability of discontinuity. If discontinuity porosity n_d is defined as the ratio of total area $\sum_{i=1}^{N}A_d^i$ of discontinuities to total area A_t (Figure 5.2b):

$$n_d = \frac{1}{A_t}\sum_{i=1}^{N}A_d^i$$

(5.41)

the following relation between apparent permeability k_{d_a} and actual permeability k_d is obtained as:

$$k_{d_a} = n_d k_d$$

(5.42)

5.3.2 Transient pulse test

In this sub-section, the finite element formulations of transient pulse tests for longitudinal and radial flow conditions are presented. Furthermore, the final forms of analytical solutions for the same tests are given.

(a) Longitudinal flow (figure 5.3)

For incremental variation δp, the integral form of Eq. (5.10) can be written as:

$$\int_{x=a}^{b}\delta p\beta\frac{\partial p}{\partial t}dx = \int_{x=a}^{b}\delta p\frac{\partial^2 p}{\partial x^2}dx$$

(5.43)

Applying integration by parts to the equation above yields the weak form of the governing equation as

$$\int_{x=a}^{b}\delta p\beta\frac{\partial p}{\partial t}dx + \int_{x=a}^{b}\frac{\partial \delta p}{\partial x}\frac{\partial p}{\partial x}dx = \delta p\hat{t}\Big|_{x=a}^{b}, \quad \hat{t}=\frac{\partial p}{\partial x}n$$

(5.44)

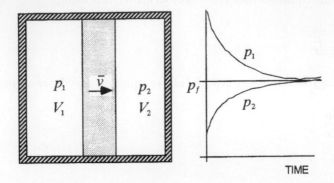

Figure 5.3 Longitudinal transient pulse test model.

Let us assume that the domain is discretized into M elements. Since Eq. (5.44) is valid for the total domain, it must also hold element-wise. Let us further assume that pressure in an element is interpolated as given below:

$$p = N_i P_i + N_j P_j \quad \text{or} \quad p = [N]\{P(t)\} \tag{5.45}$$

where

$$[N] = \begin{bmatrix} N_i & N_j \end{bmatrix}, \quad \{P\}^T = \{ P_i \quad P_j \}, \quad N_i = \frac{x_j - x}{L}, \quad N_j = \frac{x - x_i}{L}, \quad L = x_j - x_i$$

Thus Eq. (5.44) can be written for a typical element with the use of Eq. (5.45) as

$$\int_{x=x_i}^{x_j} \beta [N]^T [N] \{\dot{P}\} dx + \int_{x=x_i}^{x_j} [B]^T [B] \{P\} dx = [N]^T \{\hat{t}\} \Big|_{x=x_i}^{x_j} \tag{5.46}$$

where

$$B_i = \frac{-1}{L}, \quad B_j = \frac{1}{L}$$

The equation above may be re-written as:

$$[M]^e \{\dot{P}\}^e + [K]^e \{P\}^e = \{F\}^e \tag{5.47}$$

where

$$[M]^e = \beta \int_{x=x_i}^{x_j} [N]^T [N] dx, \quad [K]^e = \int [B]^T [B] dx, \quad \{F\}^e = [N]^T \hat{t} \Big|_{x=x_i}^{x_j}$$

Eq. (5.47) given for a typical element may be transformed to the equation below for the total domain as

$$[M]\{\dot{P}\} + [K]\{P\} = \{F\} \tag{5.48}$$

With the use of the θ method, Eq. (5.48) takes the following form for time step t_{n+1} as

$$[M^*]\{P\}_{n+1} = \{F^*\}_{n+1} \tag{5.49}$$

where

$$[M^*] = \left(\frac{1}{\Delta t}[M] + \theta[K]\right),$$

$$\{F^*\}_n = \left(\frac{1}{\Delta t}[M] - (1-\theta)[K]\right)\{P\}_n + \theta\{F\}_{n+1} + (1-\theta)\{F\}_n$$

As understood from Eq. (5.12), the boundary conditions to solve Eq. (5.49) will vary with time. With use of Darcy's law, the velocities at the ends of a sample must be continuous in relation with those of the reservoirs. Therefore, one may write the followings:

$$q_1 = -\frac{k}{\eta}\frac{1}{A}\left(\frac{\partial p}{\partial x}n\right)_{x=a}, \quad q_2 = -\frac{k}{\eta}\frac{1}{A}\left(\frac{\partial p}{\partial x}n\right)_{x=b} \tag{5.50}$$

where A is sample cross section, and $n_{x=a} = -1$, $n_{x=b} = 1$. From the Taylor expansion of Eq. (5.12) for time step t_n for the both ends of the sample, one can obtain the followings:

$$P_1^{n+1} = P_1^n - \frac{\Delta t}{C_f V_1}q_1, \quad P_2^{n+1} = P_2^n - \frac{\Delta t}{C_f V_2}q_2 \tag{5.51}$$

Thus the boundary conditions of Eq. (5.50), which change with time, can be replaced with the equivalent boundary conditions given by Eq. (5.51) and the resulting simultaneous equation system can be solved.

Aydan et al. (1997a,b) and Aydan & Üçpırtı (1997) solved Eqs. (5.10) and (5.12) using the method of elimination (Kreyszig, 1983) instead of the Laplace transformation technique employed by Brace et al. (1968) and derived the following equation for permeability of rocks:

$$k = \frac{\eta c_f L}{A}\frac{V_1 V_2}{V_1 + V_2}\ln\left(\frac{\Delta p_o}{\Delta p}\frac{V_2}{V_1 + V_2}\right)\frac{1}{t} \tag{5.52}$$

where $\Delta p = p_1 - p_f$, $\Delta p_o = p_i - p_o$. When gas is used as a permeation fluid, p_1 and p_2 are replaced with U_1 ($=p_1^2$) and U_2 ($=p_2^2$), and permeability can be calculated using the same relation given above.

If the volume of reservoir 2 (V_2) is much greater than the volume of reservoir 1 (V_1), ($V_2 >> V_1$) (for instance, outer side of specimen is open to air) p_0 ve p_f given in

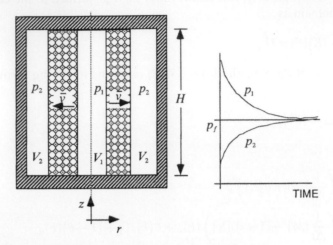

Figure 5.4 Radial transient pulse test model.

the above equation will be equal to atmospheric pressure (p_a). For this particular case, Eq. (5.52) takes the following form:

$$k = \frac{\eta c_f L V_1}{A} \ln\left(\frac{p_i - p_a}{p_1 - p_a}\right) \frac{1}{t} \tag{5.53}$$

(b) Radial flow

The finite element formulation of Eq. (5.11) for radial flow (Figure 5.4) can be obtained in a similar manner. Since the formulation would be the same except the specific forms of matrices and vectors, which will be different as a result of the geometry of the domain, it is not presented herein.

The transient pulse method is also extended to radial flow by Aydan *et al.* (1997b). The details of their method is given elsewhere (Aydan *et al.*, 1997). This method is fundamentally very similar to that for longitudinal flow (Figure 5.4). The only differences are associated with the pressure gradient and surface area at inner and outer radius. If one replaces those given in Eqs. (5.52) and (5.53) with their corresponding relations for radial flow, similar equations will be obtained. Finally, one can use the following equation to compute permeability

$$k = \eta c_f \frac{V_2 r_2 V_1 r_1 \ln(r_2/r_1)}{V_2 r_2 A_{p_1} + V_1 r_1 A_{p_2}} \ln\left(\frac{\Delta p_o}{\Delta p} \frac{V_2 r_2 A_{p_1}}{V_2 r_2 A_{p_1} + V_1 r_1 A_{p_2}}\right) \frac{1}{t} \tag{5.54}$$

When gas is used as a permeatation fluid, p_1 and p_2 are replaced with $U_1 (= p_1^2)$ and $U_2 (= p_2^2)$, and permeability can be calculated using the same relation given above.

If the volume of reservoir 2 (V_2) is much greater than the volume of reservoir 1 (V_1), $(V_2 >> V_1)$ (for instance, outer side of specimen is open to air) p_o ve p_f given in

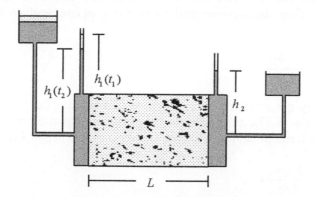

Figure 5.5 Illustration of longitudinal falling head test.

the above equation will be equal to atmospheric pressure (p_a). For this particular case, Eq. (5.54) becomes

$$k = \eta c_f \frac{V_1 r_1 \ln (r_2/r_1)}{A_{p_1}} \ln \left(\frac{p_i - p_a}{p_1 - p_a} \right) \frac{1}{t} \tag{5.55}$$

5.3.3 Falling head tests

When rock is quite permeable, falling head tests, which utilize dead weight of fluid, are also used for determining permeability of rocks and discontinuities. In this subsection, analytical solutions for falling head tests for longitudinal flow and radial flow conditions are derived.

(a) Longitudinal falling head test method

Experimental set-up used for this kind test is shown in Figure 5.5 (Aydan *et al.*, 1997b). As seen from the Figure two manometers having cross sections a are assumed to be attached to the both ends of the sample. During a test, the change of pressure and velocity of flow can be measured through these manometers. The level h_2 of water at the lower tank is assumed to be constant in the following formulation. When an experiment starts, flow rate inside the pipe can be given as:

$$v_p = -a \frac{\partial h_1}{\partial t} \tag{5.56}$$

where h_1 is the level of water inside the manometer (1). At a given time, flow rate through the cross-section area A) of the specimen is given by

$$v_t = \bar{v} A \tag{5.57}$$

It is assumed that flow rate through the specimen should be equal to the flow rate of the pipe. Then, the pressure gradient in specimen can be given in the following form:

$$\frac{\partial p}{\partial x} \approx -\rho g \frac{(h_1 - h_2)}{L} \tag{5.58}$$

where ρ is density, g is gravitational acceleration. Substituting Eq. (5.58) together with Eq. (5.40) into Eq. (5.57), and equalizing the resulting equation to Eq. (5.56) yields the following differential equation for the change of water height h_1:

$$\frac{\partial h_1}{h_1 - h_2} = -\frac{kA\rho g}{La\eta} \partial t \tag{5.59}$$

where L is sample length. Solution of the above differential equation is:

$$h_1 = h_2 + Ce^{-\alpha t} \tag{5.60}$$

where

$$\alpha = \frac{kA}{La} \frac{\rho g}{\eta}$$

If initial conditions are given by

$$h_1 = h_{10} \quad \text{at } t = 0$$

where h_{10} is water height at manometer 1 at $t = 0$. Thus the integration coefficient C is obtained as follows

$$C = h_{10} - h_2 \tag{5.61}$$

Inserting the above integration coefficient in Eq. (5.60) yields the following

$$-\alpha t = \ln\left(\frac{\Delta h}{\Delta h_o}\right) \tag{5.62}$$

where

$$\Delta h = h_1 - h_2, \quad \Delta h_o = h_{10} - h_2$$

If α is substituted into the above equation, the following expression for permeability is obtained

$$k = \frac{La}{A} \frac{\ln\left(\frac{h_{10}-h_2}{h_1-h_2}\right)}{t} \frac{\eta}{\rho g} \tag{5.63}$$

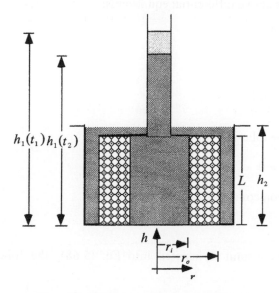

Figure 5.6 Illustration of radial free-fall test.

(b) Radial falling head test method

Experimental set-up used for this kind test is shown in Figure 5.6 (Aydan *et al.*, 1997b). As seen from the figure a manometer is placed on the top of the cylindrical hole drilled in the middle of test specimen. The cross-section area of this manometer is denoted by A_h. During the test, the change of pressure and velocity of flow can be measured through this manometer. The level h_2 of water at the outer container is assumed to be constant. When the experiment starts, flow rate inside the manometer can be given as:

$$q = -\rho g A_h \frac{\partial h_1}{\partial t} \tag{5.64}$$

where h_1 is the level of water inside the manometer. At a given time, the flow rate through a cross-section area of hole (A_p) inside test specimen is given by

$$v_t = \bar{v} A_p \tag{5.65}$$

It is assumed that flow rate through the hole perimetry should be equal to the flow rate of the pipe. The pressure gradient in specimen may be given in the following form:

$$\frac{\partial p}{\partial r} \approx -\frac{\partial}{\partial r}(\rho g(h_1 - h_2)) = -\rho g \frac{\partial (h_1 - h_2)}{\partial r} = -\rho g \frac{(h_1 - h_2)}{r \ln (r_o/r_i)} \tag{5.66}$$

Substituting Eq. (5.66) into Eq. (5.65), and equalizing the resulting equation to Eq. (5.64) yields the following differential equation for the change of water height h_1:

$$\frac{\partial h_1}{h_1 - h_2} = -\frac{k}{\eta} \frac{A_p}{A_h} \frac{1}{r_i \ln (r_o/r_i)} \partial t \tag{5.67}$$

Solution of the above differential equation is:

$$h_1 = h_2 + Ce^{-\alpha t} \tag{5.68}$$

where

$$\alpha = \frac{k}{r_i \ln(r_o/r_i)} \frac{A_p}{A_h} \frac{1}{\eta}$$

Introducing the following initial conditions

$$h_1 = h_{10} \quad \text{at } t = 0$$

yields the integration constant C as:

$$C = h_{10} - h_2 \tag{5.69}$$

If integration constant is inserted into Eq. (5.68), the following equation is obtained:

$$-\alpha t = \ln\left(\frac{h_1 - h_2}{h_{10} - h_2}\right) \tag{5.70}$$

If α is substituted into the above equation, the following expression for permeability is obtained

$$k = \eta r_i \ln\left(\frac{r_o}{r_i}\right) \frac{A_h}{A_p} \ln\left(\frac{h_{10} - h_2}{h_1 - h_2}\right) \frac{1}{t} \tag{5.71}$$

5.4   SOME SPECIFIC SIMULATIONS AND APPLICATIONS TO ACTUAL EXPERIMENTS

5.4.1   Some specific simulations

In the first part of this section, the following two cases are analysed using the finite element method presented in the preceding section:

- *Case 1:* The compressibility of rock and fluid are constant, and porosity is variable
- *Case 2:* The compressibility of rock is variable, and the compressibility of fluid and porosity are constant

(a) The effect of porosity

In order to investigate the effect of porosity, the compressibility of solid phase and fluid phase is assumed to be constant while porosity is varied. The effect of variation of porosity enters the equation system through constant β. Bulk compressibility C_b is generally obtained from experiments. We can also make estimation from the limiting values of the bulk compressibility from the compressibility of a sample when porosity

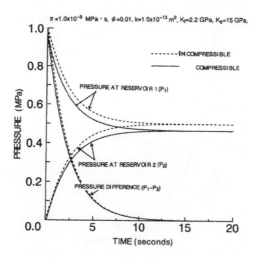

Figure 5.7 Variation of pressure difference with time (porosity $\phi = 0.01$).

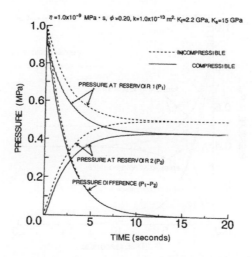

Figure 5.8 Variation of pressure difference with time (porosity $\phi = 0.20$).

has values of 0 and 1. When the porosity has a value of 0, the bulk compressibility must be equal to that of solid phase. On the other hand, when the porosity has a value of 1, then the bulk compressibility must be equal to that of fluid phase. From these requirements it simply implies that the bulk compressibility is twice that of solid phase. In each computation, the value of constant β is computed using this requirement. Figures 5.7 and 5.8 show computed results.

In the same figures, relations used in interpreting transient pulse tests are also included for comparisons. As seen from these figures, the computed relations become markedly different from the conventionally used relations as the value of porosity increases. From this observation it can be said the relations used in interpreting

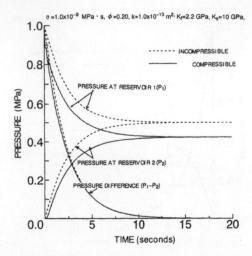

Figure 5.9 Variation of pressure difference with time (compressibility of solid phase $C_s = 0.1 \, GPa^{-1}$).

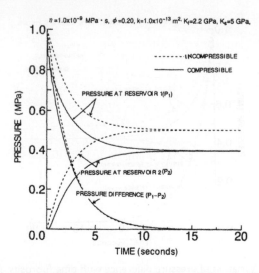

Figure 5.10 Variation of pressure difference with time (compressibility of solid phase $C_s = 0.2 \, GPa^{-1}$).

transient pulse test results may be only valid when porosity is very small. Otherwise, the estimated permeability values will be greater than the actual ones.

(b) The effect of compressibility of solid phase

In order to investigate the effect of the compressibility of solid phase, the compressibility of fluid phase and porosity are assumed to be constant while the compressibility of solid phase is varied. The value of constant β is computed using the procedure described in the previous case. Figures 5.9 and 5.10 show computed results. In the same figures, relations used in interpreting transient pulse tests are also included for comparisons.

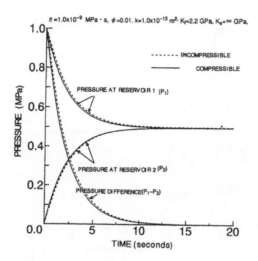

Figure 5.11 Variation of pressure difference with time (compressibility of solid phase $C_s = 0.00\,GPa^{-1}$).

As also seen from these figures, the computed relations become markedly different from the conventionally used relations as the value of solid phase compressibility increases. From this observation it can also be said the relations used in interpreting transient pulse test results are only valid when porosity and the compressibility of solid phase are very small. Otherwise, the estimated permeability values will be greater than the actual ones.

In the next series of computations, the effect of the rigidity of solid phase is investigated when and the value of porosity is small. The compressibility of solid phase was assumed to have the values of 0.0 and 0.02 and the value of porosity was taken as 0.01. The value of constant β is computed using the procedure described in the previous case. Figures 5.11 and 5.12 show computed results. In the same figures, relations used in interpreting transient pulse tests are also included for comparisons. As also seen from these figures, the computed relations become very close to the conventionally used relations as the solid phase becomes rigid.

5.4.2 Applications to actual permeability tests

Figure 5.13 shows pressure responses of Reservoirs 1 & 2 in a transient pulse test on a sandstone sample. Despite some scattering of pressure responses the variations of reservoir pressures tend to decrease with time and become asymptotic to a stabilizing pressure. The permeability of the sandstone sample was $3.1 \times 10^{-12}\,m^2$.

Üçpırtı & Aydan (1997) and Üçpırtı et al. (1992) reported transient pulse type experiments on halite rock sample and halite-concrete interface as a part of an experimental study on the permeability of interface between sealing plug and surrounding rock in the safe disposal of nuclear wastes. In experiments, the permeability of rock and interfaces with or without a polyethylene membrane is measured under various loading conditions to simulate possible stress state in-situ. From this experimental study,

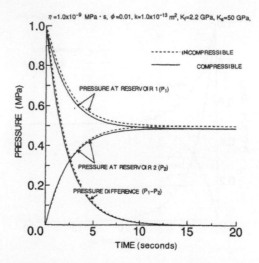

Figure 5.12 Variation of pressure difference with time (compressibility of solid phase $C_s = 0.02$ GPa^{-1}).

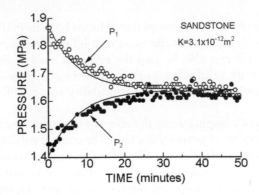

Figure 5.13 Pressure responses of a sandstone sample in a transient pulse test.

it is concluded that the membrane could be very effective in reducing the permeability of the interface which is the most likely path of flow within the system.

An in-situ application of the falling-head radial flow experiment for permeability (hydraulic conductivity) measurements was carried out at Gündoğdu district of Babadağ town in Denizli province of Turkey by Aydan *et al.* (2003). The ground in the permeability and infiltration site consists of intercalated marl and loosely cemented sandstone (Figure 5.14). Three boreholes with a diameter of 90 mm were drilled and they were separated from each other by a distance of about 1050 mm. Thus the horizontal distance between the boreholes were about 10 times the borehole diameter. The depth of boreholes numbered SK-1, SK-2, SK-3 was 1260, 1300 and 1420 mm, respectively. The boreholes were equipped with water pressure sensors and the water pressure was measured during a test using a WE7000 (YOKOGAWA) data acquisition system and data were stored onto a laptop computer.

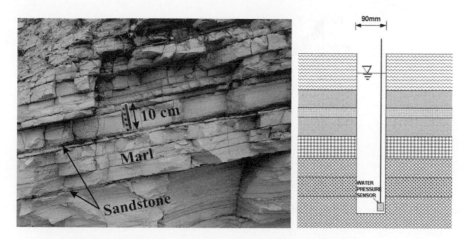

Figure 5.14 A view of intercalated marl and loosely cemented sandstone at test site and sketch of in-situ permeability test.

Four permeability tests at two boreholes were carried out simultaneously. For this particular problem there are several semi-empirical formulas to compute the hydraulic conductivity of ground. One of the formula is purposed by Barron *et al.* (1970), which has the following form

$$k = 2.3a \log_{10}\left[\frac{H_1}{4H_2} \cdot \frac{1}{t_2 - t_1}\right] \tag{5.72}$$

where H_1, H_2 and a are water heads at time t_1 and t_2 ($t_2 > t_1$ and $H_1 > H_2$) and radius of the borehole.

The problem of permeability tests in this particular case may be contemplated as an axi-symmetric problem and the governing equation for radial flow takes the following form:

$$\frac{\partial h}{\partial t} = -k\frac{1}{r}\left(r\frac{\partial h}{\partial r}\right) \tag{5.73}$$

where h is water head and r is radial distance. Finite Element Method was adopted to solve Eq. (5.73). The finite element formulation of Eq. (5.73) and its solution follows the same procedure as used in the transient pulse method experiments. Figure 5.15 compares the computational results with measured data together with estimated response from the analytical solution for falling-head tests. The computed hydraulic conductivity coefficients according to Barron's formula and FEM are given Table 5.1. As seen from Table 5.1 and Figure 5.15, the computed and measured results are very close to each other.

Permeability experiments were carried out on granite samples containing 1 and 2 discontinuity planes using falling-head experimental technique. The discontinuity spacing was 200 mm. The inner diameter of the hole in the center of the specimen

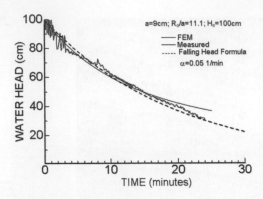

Figure 5.15 Comparison of measured water head variation with those estimated from FEM and falling-head test formula.

Table 5.1 Inferred hydraulic conductivities of tests (from Kumsar et al., 2016).

Test No	Hydraulic Conductivity (cm/s) (Barron's formula)	Hydraulic Conductivity (cm/s) (FEM)
Test 1–1	5.482×10^{-4}	6.815×10^{-4}
Test 1–2	1.250×10^{-3}	1.356×10^{-3}
Test 2–1	3.523×10^{-4}	2.521×10^{-4}
Test 2–2	5.000×10^{-4}	6.560×10^{-4}
Test 3–1	3.584×10^{-4}	2.568×10^{-4}
Test 3–2	5.701×10^{-4}	7.640×10^{-4}
Test 4–1	3.523×10^{-4}	2.503×10^{-4}
Test 4–2	4.710×10^{-4}	4.936×10^{-4}
Test 5	4.727×10^{-4}	4.831×10^{-4}
Test 6	6.686×10^{-4}	4.942×10^{-4}

was 64 mm. The side length of the granite blocks was 200 mm. Figure 5.16 shows the measured response of water-head in the central borehole with time. The water head was almost nil after a certain period of time (30–50 seconds). The value of coefficient α given by Eq. (5.68) ranged between 1/10–1/12 seconds after the curve-fitting to the experimental data.

Figure 5.17 shows the measured response of water-head in the central borehole with time for the sample having two discontinuity planes with the spacing of 200 mm. The water head was almost nil after a certain period of time (30–50 seconds). The value of coefficient α given by Eq. (5.68) ranged between 1/11–1/12 seconds after the curve-fitting to the experimental data. The value of the constant (α) was quite similar to the sample having a single discontinuity.

If the total water flow is assumed to be through the discontinuity plane/planes, the hydraulic aperture of the discontinuity plane can be obtained from a similar formulation given in Sub-sections 5.3.1 and 5.3.3 with due consideration of the axi-symmetry condition of the experiments (see Aydan *et al.*, 1997a,b for details).

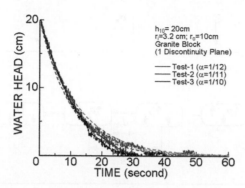

Figure 5.16 Experimental water head variations with time having a single discontinuity plane in a granite block.

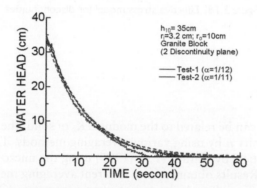

Figure 5.17 Experimental water head variations with time having two discontinuity planes in a granite block.

5.5 MECHANICAL COUPLING EFFECT OF GROUNDWATER ON ROCKS AND DISCONTINUITIES

5.5.1 Theoretical formulation

Terzaghi (1925) has proposed an effective stress law which is defined as:

$$\sigma'_{ij} = \sigma_{ij} - p\delta_{ij} \tag{5.74}$$

where σ_{ij} is the total stress tensor and p is the fluid pressure (compression is assumed positive), and δ_{ij} is Kronecker delta. This concept is later also used for rocks. On the other hand, Biot (1942) introduced a parameter α which is related to the volumetric porosity n and the ratio of volumetric stiffness K_s of solid and that of the bulk as:

$$\sigma'_{ij} = \sigma_{ij} - \alpha p\delta_{ij} \tag{5.75}$$

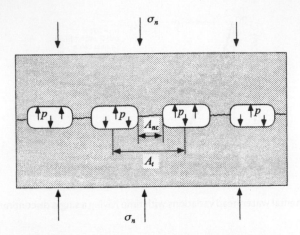

Figure 5.18 Effective stress model for discontinuities.

where

$$\alpha = 1 - \frac{K}{K_s}$$

Bulk modulus K can be related to the modulus K_s of solid, the modulus K_f of fluid and volumetric porosity n by using various averaging methods. These approaches are summarised in an article by Karaca *et al.* (1995) using the micro-structure models of Aydan *et al.* (1992). Results obtained from different averaging methods imply that the value of α can range from 0 to 1, depending upon parameters K_s, K_f and n. It seems that Terzaghi-type effective stress law is a special case among various models.

Terzaghi's and Biot's model were also extended to account the mechanical effect of water on discontinuities in rock masses. Byerlee (1967) was the first to suggest an effective stress model by employing a Terzaghi-type effective stress concept. The stress tensor and volumetric porosity n are replaced by normal stress σ_n and the ratio η of the non-contact area A_{nc} to the total area A_t between the discontinuity walls in the following forms (Figure 5.18):

$$\sigma_n' = \sigma_n - \alpha p, \quad \eta = \frac{A_{nc}}{A_t}, \quad \alpha = \left(1 - \frac{E_f}{E_s}\right) \quad E_f \approx \frac{K_f}{3} \tag{5.76}$$

Similarly, the value of α can have a value ranging from 0 to 1, depending upon the chosen mechanical model. The value of α has a large variation and its value is still not well-understood. In the literature, it is very rare to see any discussion about this matter and many people seems to follow the approach proposed by either Terzaghi or Biot.

5.5.2 Theoretical modelling of tilting tests

Biot suggested a genious yet simple experimental set-up, which is known as the *Beer-Can experiment* to investigate the mechanical effect of water on the sliding resistance

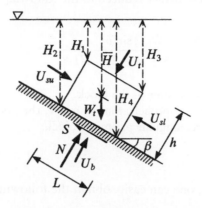

Figure 5.19 Mechanical model for prismatic block.

of rock discontinuities (quoted in an article by Hubbert & Rubey 1959). The author devised experimental set-ups which are slightly modified from the original experimental set-up suggested by Biot. Theoretical formulations for tilting tests on rectangular prismatic blocks and wedges are presented herein.

(a) Theoretical formulations of tilting test set-up for rectangular prismatic blocks

Let us consider a prismatic block put on plane β, submerged in a fluid (Figure 5.19). Force equilibrium equations for the block can be written as

$$\sum F_s = W_t \sin \beta + U_{su} - U_{sl} - S = 0 \tag{5.77a}$$

$$\sum F_n = N + \alpha U_b - W_t \cos \beta - U_t = 0 \tag{5.77b}$$

where

$$W_t = (1 - n) W_{br} + n W_{bw}, \quad W_{bw} = \gamma_w h L B,$$

$$U_{sl} = \gamma_w \frac{H_3 + H_4}{2} h B, \quad U_{su} = \gamma_w \frac{H_1 + H_2}{2} h B,$$

$$U_t = \gamma_w \frac{H_1 + H_3}{2} L B, \quad U_b = \gamma_w \frac{H_3 + H_4}{2} h B,$$

Let us also assume that the block at a limiting equilibrium state such that

$$S = \mu_s N, \quad \mu_s = \tan \phi_s, \tag{5.78}$$

Using the above relations, the following relations can be obtained:

$$\mu_s = \frac{W_t \sin \beta - U_{sl} + U_{su}}{W_t \cos \beta - \alpha U_b + U_t} \tag{5.79}$$

This expression can be further reduced to the following form using the geometry and above relations as:

$$\mu_s = \frac{(1-n)(W_{br} - W_{bw})\sin\beta}{\left[W_t - \frac{1+3\alpha}{4}W_{bw}\right]\cos\beta + (\alpha-1)\gamma_w\overline{H}LB} \tag{5.80}$$

where $\overline{H}$ is the depth of the center of the block.

If we follow Terzaghi's approach, that is, $\alpha = 1$, the above expression reduces to the following form for a rectangular prismatic block:

$$\mu_s = \tan\beta \tag{5.81}$$

Under dry condition, one can easily obtain the following relation using a similar kind formulation:

$$\mu_d = \tan\beta \tag{5.82}$$

If the friction angles obtained from tilting tests under dry and submerged conditions are same, that is $\mu_d = \mu_s$, then it will imply that the effective stress law of Terzaghi-type should hold for rock discontinuities.

(b) Theoretical formulations of tilting test set-up for wedge blocks

Let us consider a wedge-like block as shown in Figure 5.20. Force equilibrium equations for each respective direction can be given by

$$\sum F_s = W\sin\psi - U_s\cos\psi + U_t\sin\psi - S = 0 \tag{5.83a}$$

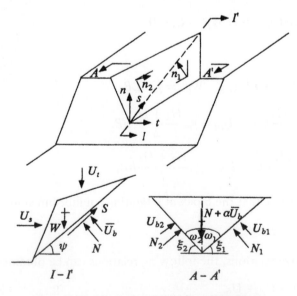

Figure 5.20 Mechanical model for wedge block.

$$\sum F_n = W \cos \psi + U_s \sin \psi + U_t \cos \psi - N - \alpha \overline{U}_b = 0 \tag{5.83b}$$

$$\sum F_t = -(N_1 + \alpha \overline{U}_{b1}) \cos \omega_1 + (N_2 + \alpha U_{b2}) \cos \omega_2 = 0 \tag{5.83c}$$

where $N + \alpha \overline{U}_b = (N_1 + \alpha U_{b1}) \sin \omega_1 + (N_2 + \alpha U_{b2}) \sin \omega_2$, ψ is intersection angle. $W = (1-n)W_{br} + nW_{bw}$. Note that In Eqns. 5.83b and 5.83c, shear force components perpendicular to intersection line on planes ω_1 and ω_2 are neglected since the motion of block perpendicular to the intersection line direction is negligible.

Let us assume that the failure planes have no cohesion and obey the simple failure criterion as given below:

$$T = (N_1 + N_2)\mu; \quad \mu = \tan \phi \tag{5.84}$$

and introducing a safety factor as:

$$SF = \frac{T}{S} \tag{5.85}$$

With the use of Eqs. (5.83) and (5.84), the above expression yields the following:

$$SF = \frac{[(W \cos \psi + U_s \sin \psi + U_t \cos \psi) \cdot \lambda - \alpha U_b^*]}{W \sin \psi - U_s \cos \psi + U_t \sin \psi} \mu \tag{5.86}$$

where $U_b^* = U_{b1} + U_{b2}$. λ is termed as the wedge factor given by (Kovari & Fritz, 1975)

$$\lambda = \frac{\cos \omega_1 + \cos \omega_2}{\sin (\omega_1 + \omega_2)}$$

Assuming that $U_s, U_t, U_{b1}, U_{b2} = 0$, which corresponds to *dry case*, together with $SF = 1$ yields the apparent friction angle ϕ^* due to the geometric configuration of the wedge is obtained as:

$$\phi_d^* = \tan^{-1}(\lambda \tan \phi) \tag{5.87}$$

The maximum wedging effect is obtained when $\omega_1 = \omega_2 = \omega$.

If α is chosen as 1, which corresponds to Terzaghi type effective stress law, and the geometry of the block is considered, the resulting equation takes the following form

$$\phi_s^* = \tan^{-1}(\lambda \tan \phi) \tag{5.88}$$

This implies that if Terzaghi type effective stress holds, the apparent friction angle of the wedge should be the same both under dry and submerged conditions unless there is a chemical reaction between rock and water along sliding planes.

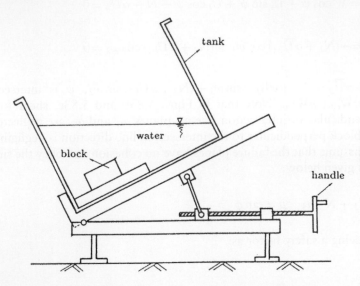

Figure 5.21 Experimental setup under submerged condition.

5.5.3 Tilting experiments

An experimental set-up was designed as shown in Figure 5.21. The device consists of a tilting apparatus with an inclinometer and a mountable water tank. The tests on discontinuities are carried out by mating walls of prismatic rock blocks and model wedges with fixed base blocks of the same material under dry and submerged conditions.

(a) Tests on rectangular prismatic blocks

During some tests, the water head is varied since the Biot-type formulation depends upon the water head $\overline{H}$, porosity n of intact rock, and the coefficient α. The discontinuities tested were either artificially made or natural. Figure 5.22 shows the effect of water $\overline{H}$ on various kind discontinuities. As seen from the plots, the friction angle under submerged condition seems not to be influenced by the variation of water head at all. The maximum variation is restricted to 2–3%. It is also interesting to note that there is not any remarkable change in the friction angle under fully submerged condition.

Figure 5.23 shows the plots of a number of test results on artificial and natural discontinuities in various kinds of rock. The Figure is a plot of dry friction angles versus to those under fully submerged conditions. It is interesting to note that the experimental results are either on or both sides of the line for $\alpha = 1$. The maximum variation on both sides of the line is limited to ±3%.

5.5.4 Tests on wedge blocks

Six special moulds were prepared to cast model wedges. For each wedge configuration, three wedge blocks were prepared. Each base block had dimensions of

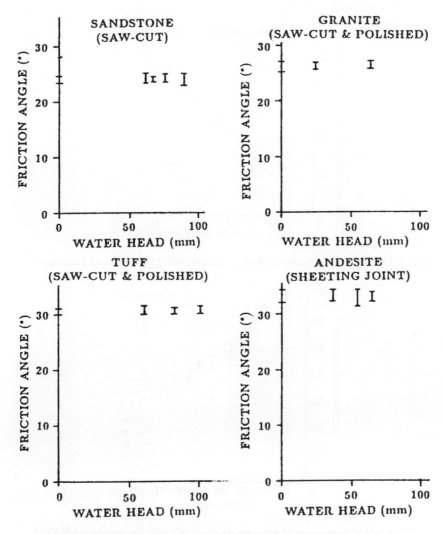

Figure 5.22 Effect of water head on the friction angle of discontinuities.

$140 \times 140 \times 260$ mm and made of mortar (Kumsar *et al.*, 1997). The cement used in mortar was rapid hardening type and samples were cured for about 7 days in a room with a constant temperature. In addition, several mortar slabs were cast to measure the friction angle of sliding planes. The mean friction angle of sliding planes was 35° with a standard deviation of ±2°.

The model wedge blocks were all stable when they were set on their base blocks. A tilting device was used to cause the sliding of wedge blocks. The intersection angle of blocks were increased until the sliding of the block occurs. For each wedge block three tests were carried out.

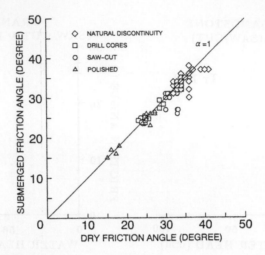

Figure 5.23 Relationship between dry and submerged friction angles of discontinuities.

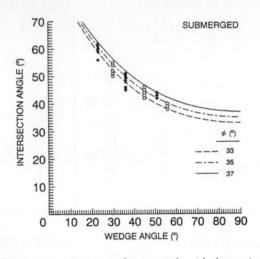

Figure 5.24 Comparison of apparent friction angle with theoretical predictions.

Tests results together with theoretical predictions for friction angles of 33°, 35° and 37° by assuming that $\alpha = 1$ are shown in Figure 5.24. For each wedge geometry, 9 experimental results are plotted in this figure. As seen from this figure, the experimental results closely follow the theoretical curves, although some scattering exists. This scattering was considered to have been caused by the use of fine sand in mortar which results in some stick-slip behaviour of sliding planes and a variation of the frictional characteristics of the planes due to penetration of sand particles into the base block.

(b) Direct shear tests

Direct shear tests on discontinuities of granitic rock with polished surfaces were carried out to see the mechanical effect of water on the shear resistances of rock discontinuities

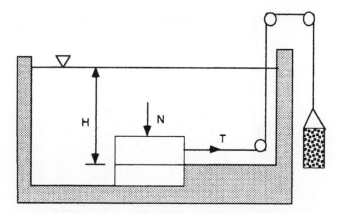

Figure 5.25 Experimental set-up for direct shear tests.

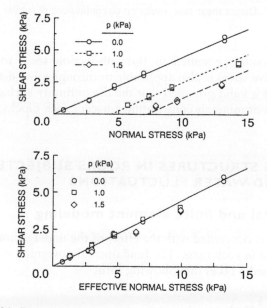

Figure 5.26 Direct shear test results on discontinuities of granitic rocks.

by using an experimental set-up as shown in Figure 5.25. Tests were carried out on dry samples and submerged samples by varying the water head and plotted in Figure 5.26 using Terzaghi's effective normal stress as $\sigma_{n'} = \sigma_n - p$. As seen from the figure, tests results seem to confirm the validity of the effective stress definition of Terzaghi-type for rock discontinuities.

Additional direct shear tests were also performed on discontinuities of rubber material with flat and undulating surfaces of sinusoidal-type. Figure 5.27 shows the experimental results by using Terzaghi's effective stress law. Once again, these tests results also confirm the validity of the above conclusions.

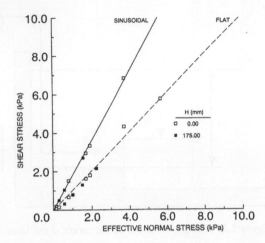

Figure 5.27 Direct shear test results on discontinuities of rubber material.

Experimental results presented in this sub-section seem to confirm that the Terzaghi-type effective stress is also applicable to throughgoing discontinuities. However, this conclusion is valid provided that the discontinuity surfaces are not affected by water absorption of minerals constituting adjacent rock blocks.

5.6 MODELING STRUCTURES IN ROCKS SUBJECTED TO GROUND-WATER FLUCTUATIONS

5.6.1 Theoretical and finite element modeling

The first analysis was concerned with the effect of the underground water on ground water table variation in rock mass. The fundamental governing equation (Eq. 5.8) of seepage in porous media takes the following form:

$$\frac{n}{K_f}\frac{\partial p}{\partial t} = -\nabla \cdot \left(\frac{k}{\eta}\nabla p\right) \tag{5.89}$$

where n, K_f, k, η and p are porosity, fluid compressibility, permeability and pore pressure respectively. Pore pressure is related to water head in the following form

$$p = \rho_f g h \tag{5.90}$$

where ρ_f, g and h are fluid density, gravity and water head, respectively. Using Eq. (5.90) in Eq. (5.89), one can easily get the following equation

$$S\frac{\partial h}{\partial t} = -k\nabla \cdot \nabla h \tag{5.91}$$

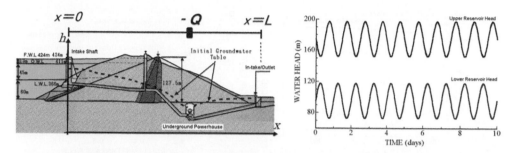

Figure 5.28 The boundary and initial conditions of one-dimensional finite element model for seepage analysis.

where S is storage coefficient. Using the concept of Biot (1946), effective stress is generally written in the following form:

$$\sigma' = \sigma - \alpha p \mathbf{I} \tag{5.92}$$

where α is Biot coefficient. When Terzaghi's effective stress concept is adopted, its value becomes 1.

The equation of motion can be written in incremental form in the following form:

$$\nabla \cdot \dot{\sigma} = 0 \tag{5.93}$$

If one follows the conventional form of formulation based on the finite element method, governing equations (5.91) and (5.93) takes the following form:

$$[M]\{\dot{H}\} + [G]\{H\} = \{R\} \tag{5.94}$$

$$[K]\{\dot{U}\} = \{\dot{F}\} \tag{5.95}$$

where

$$[M] = S \int_V [L]^T [L] dV; \quad [G] = k \int_V [A]^T [A] dV; \quad [A] = \nabla[L];$$

$$\{R\} = \int_\Gamma [\bar{L}] q_n d\Gamma + \int_V [L] Q dV; \quad [K] = \int_V [B]^T [D][B] dV; \quad \{\dot{F}\} = \int_V [B]^T \{\dot{P}\} dV$$

5.6.2 Applications to pumped storage power house project

The problem is first treated as a one-dimensional problem with an emphasis on water head variation. The water heads at upper and lower reservoirs were changed as shown in Figure 5.28. The powerhouse was considered as a sink source and the initial water head condition was assigned using Dupuit's formula (i.e. Verruijt, 1982). Figures 5.29 and 5.30 show the water head variations at selected points at chosen time steps for the underground powerhouse with impermeable lining or no lining, respectively. When

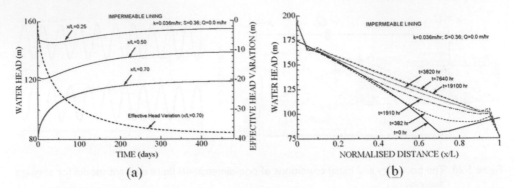

Figure 5.29 Variation of water head at selected points or time (impermeable lining).

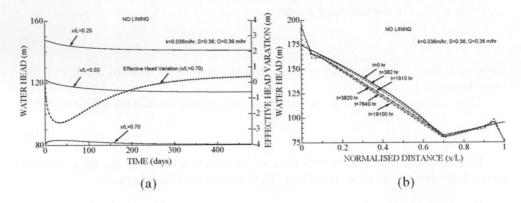

Figure 5.30 Variation of water head at selected points or time (no lining).

the lining is impermeable the water head variation reaches steady state after about 300 days and the variation of the water head in rock mass due to the variation of water level of reservoirs is limited to 60–70 m from the intakes/outlets. There is almost no change on the water head in the vicinity of the powerhouse. When the cavern is unlined, the water head variations are much smaller and they are limited to 60–70 m from the intakes/outlets. However, the steady state is achieved after about 400 days.

Next computation was carried using a two-dimensional coupled seepage analyses and the mesh used is shown in Figure 5.31. The water head variations were assumed to be the same as those in the previous case and the cavern was unlined. Figure 5.32 shows the water head distribution and displacement vectors when the water is drawn from upper reservoir. As noted from the figure, the rock mass below the upper reservoir rebounds while the rock mass below the lower reservoir subsides. However, when the upper reservoir is re-filled, it tends to return its original state while the rock mass below the lower reservoir rebounds as seen in Figure 5.33. This process repeats itself as the water levels of reservoirs are cyclically varied. The large variations of water heads in rock mass occur in the close vicinity of the reservoirs and their effect is quite negligible

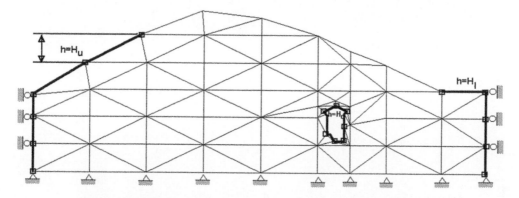

Figure 5.31 Finite element mesh for coupled hydro-mechanical analyses.

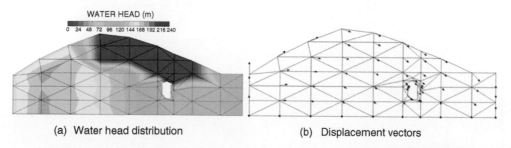

(a) Water head distribution   (b) Displacement vectors

Figure 5.32 Computed responses for 45 m water level decrease of the upper reservoir.

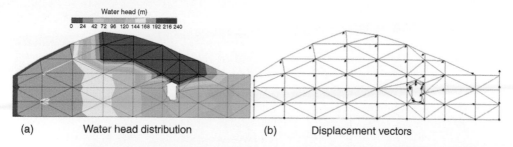

(a)   Water head distribution   (b)   Displacement vectors

Figure 5.33 Computed responses for 45 m refill of the upper reservoir.

around the cavern. This conclusion is quite similar to that from the one-dimensional finite element analyses. These results further imply that there is no need to take into account the effective stress changes around the caverns of the pumped storage schemes in long term. Nevertheless, they may have important implications on the slopes of the reservoirs due to large variations of effective stresses.

Chapter 6

Thermo-hydro-diffusion behaviour of rocks

6.1 INTRODUCTION

The coupled problems are commonly encountered in geo-science and geo-engineering. The safe disposal of nuclear wastes, petroleum and geothermal exploitations are some of examples for such problems in geo-engineering. Similar problems are also encountered in earthquake prediction studies in geo-science. The radioactive nuclear waste disposal particularly receives great attention and one may find many studies in literature. Most of these studies on radioactive nuclear waste disposal are based on thermo-hydro-mechanical concepts and it is very rare to see any study, which includes diffusion phenomena of radioactive substances. Therefore, a mechanical model, which includes 4 different aspects, should be the most appropriate approach for radioactive waste disposal. Although the variation of stress field around the disposal site may be of great concern for cavity stability in very near field, its effect is expected to be quite limited after back-filling the excavated space in long term. In this article, the mechanical effect is omitted in the coupling model presented herein and a theoretical formulation, based on the mixture theory is described for the thermo-hydro-diffusion phenomena. In the theoretical formulation, Duffour and Soret effects are considered for coupling the thermal and diffusion fields with each other. Then, a finite element formulation of the coupled model is presented and it is used for numerical analysis of some laboratory tests and compared with experimental results. Furthermore, a series of parametric numerical analyses are performed to investigate

a) Simulation of solute transport in rock under laboratory conditions,
b) Temperature field of geo-thermal fields under forced seepage
c) Non-isothermal advective moisture transport through buffer materials, and

Parametric studies on the consideration of Duffour and Soret laws in purely coupled hydro-thermo-diffusion problems.

6.2 MECHANICAL MODELING

The mechanical modeling of thermo-hydro-diffusion phenomena is based on the mixture theory (Trusdell & Toupin, 1960; Eringen & Ingram, 1965). Figure 6.1 illustrates how three fields are coupled. Following the principles of the mixture theory, and some

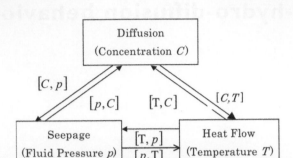

Figure 6.1   Coupling model.
[C,p]: Effect of Diffusion on Seepage (density variation)
[p,C]: Effect of Seepage on Diffusion (Advection of Concentration due to fluid flow)
[T,p]: Effect of Heat Flow on Seepage (Density variation due to Temperature variation)
[p,T]: Effect of Seepage on Heat flow (Advection of Temperature due to fluid flow)
[T,C]: Effect of Heat Flow on Concentration (Duffour effect)
[C,T]: Effect of Diffusion on thermal field (Soret effect)

appropriate constitutive relations, one can easily derive that the fundamental equations for each field as follows (Aydan, 2001).

6.2.1   Fundamental equations

The mechanical modelling of thermo-hydro-diffusion phenomena is based on the mixture theory (Trusdell & Toupin, 1960; Eringen & Ingram, 1965). Figure 6.1 illustrates how three fields are coupled.

Following the principles of the mixture theory, one can easily derive that the fundamental equations for each field are as follows (Aydan, 2001a,b; Bear, 1988):

Seepage field

$$\frac{1-n}{\rho_s}\frac{\partial \rho_s}{\partial t} + \frac{n}{\rho_f}\frac{\partial \rho_f}{\partial t} + \frac{1}{\rho_s}\nabla \cdot \{\mathbf{q}_s\} + \frac{1}{\rho_f}\nabla \cdot \{\mathbf{q}_f\} = 0 \tag{6.1}$$

where $\mathbf{q}_s = (1-n)\,\rho_s\mathbf{v}_s$ and $\mathbf{q}_f = n\rho_f\mathbf{v}_f$.

Diffusion field

$$\frac{\partial C}{\partial t} + \mathbf{v}_s \cdot \nabla\{(1-n)C_s\} + \mathbf{v}_f \cdot \nabla\left(nC_f\right) = -\nabla \cdot \{\mathbf{f}\} \tag{6.2}$$

where $C = (1-n)\,C_s + nC_f$ and $\mathbf{f} = (1-n)\mathbf{f}_s + n\mathbf{f}_f$.

Thermal field

$$(1-n)\,\rho_s\frac{d_sU_s}{dt} + n\rho_f\frac{d_fU_f}{dt} = \nabla \cdot \mathbf{h} + Q \tag{6.3}$$

where $\mathbf{h} = (1-n)\mathbf{h}_s + n\mathbf{h}_f$ and $Q = (1-n)\,Q_s + nQ_f$.

6.2.2 Constitutive laws

Seepage field

D'Arcy law is usually used to relate the relative average fluid velocity v_r to pressure in the following form

$$v_r = n(v_f - v_s) = -\frac{k}{\eta}\nabla p \tag{6.4}$$

where k is permeability and η is viscosity coefficient of fluid. If fluid density is assumed to be a function of p, T, C, its variation may be written in the following form

$$\frac{1}{\rho_f}\frac{\partial \rho_f}{\partial t} = \frac{1}{K_f}\frac{\partial p}{\partial t} - \beta_{fT}\frac{\partial T}{\partial t} - \beta_{fC}\frac{\partial C_f}{\partial t} \tag{6.5}$$

where K_f is fluid compressibility, β_{fT} is thermal expansion coefficient and β_{fC} is diffusive expansion coefficient.

Diffusion field

Fick's law is often employed as a constitutive law for diffusion problems. This law is extended by associating concentration flux f_s of solid phase with gradients of temperature and concentration as given below:

$$f_s = -D_{sT}\nabla T_s - D_{sC}\nabla C_s \tag{6.6}$$

where D_{sT} is thermal diffusivity of solid phase and D_{sC} is Dufour's coefficient (Bear, 1988). Similarly, one may also write the following relation for fluid phase.

$$f_f = -D_{fT}\nabla T_f - D_{fC}\nabla C_f \tag{6.7}$$

If $T_s = T_f = T$ and $C_s = C_f = C$, the average concentration flux f takes the following form

$$f = -D_T\nabla T - D_C\nabla C \tag{6.8}$$

where

$$D_T = (1-n)D_{sT} + nD_{fT} \quad \text{and} \quad D_C = (1-n)D_{sC} + nD_{fC}.$$

D_C is average diffusion coefficient, D_{sC} and D_{fC} are the diffusion coefficient of solid phase and fluid phase, respectively.

Thermal field

Fourier's law is well known as a constitutive law for associating heat flux to temperature gradient. This law is also expanded by associating heat flux h_s with temperature and concentration gradients as

$$h_s = -\lambda_{sT}\nabla T_s - \lambda_{sC}\nabla C_s \tag{6.9}$$

where λ_{sT} is thermal conductivity coefficient and λ_{sC} is Soret's coefficient of solid phase. Similarly the following relation can be written for fluid phase.

$$\mathbf{h}_f = -\lambda_{fT}\nabla T_f - \lambda_{fC}\nabla C_f \tag{6.10}$$

where λ_{fT} is thermal conductivity coefficient and λ_{fC} is Soret's coefficient of fluid phase. If $T_s = T_f = T$ and $C_s = C_f = C$, then average heat flux $\mathbf{h}$ takes the following form.

$$\mathbf{h} = -\lambda_T\nabla T - \lambda_C\nabla C \tag{6.11}$$

where $\lambda_T = (1-n)\lambda_{sT} + n\lambda_{fT}$ and $\lambda_C = (1-n)\lambda_{sC} + n\lambda_{fC}$. λ_T is average conductivity coefficient. λ_{sT} and λ_{fT} are conductivity of solid and fluid phases, respectively.

6.2.3 Simplified form of fundamental equations

Seepage field

If the density variation of solid phase and the porosity variation of skeleton are negligible, Eq. (6.1) may be re-written together with D'Arcy's law as:

$$n\left(\frac{1}{K_f}\frac{\partial p}{\partial t} - \beta_{fT}\frac{\partial T}{\partial t} - \beta_{fC}\frac{\partial C}{\partial t}\right) = -\nabla \cdot \left(\frac{k}{\eta}\nabla p\right) \tag{6.12}$$

Diffusion field

Eq. (6.2) may take the following form together with the use of Eq. (6.8)

$$\frac{\partial C}{\partial t} = \nabla \cdot (D_C\nabla C + D_T\nabla T) - \mathbf{v}_r \cdot \nabla C \tag{6.13}$$

The second term on the right-hand side is the advective term.

Thermal field

Internal energies of solid and fluid phases may be related to temperature field with the use of specific heat coefficients c_s and c_f as

Solid Phase

$$\frac{\partial U_s}{\partial t} = \frac{\partial U_s}{\partial T}\frac{\partial T}{\partial t} = c_s\frac{\partial T}{\partial t} \tag{6.14}$$

Fluid Phase

$$\frac{\partial U_f}{\partial t} = \frac{\partial U_f}{\partial T}\frac{\partial T}{\partial t} = c_f\frac{\partial T}{\partial t} \tag{6.15}$$

With the use of equations above, Eq. (6.11) and $\mathbf{v}_s = 0$, Eq. (6.3) becomes

$$\rho c\frac{\partial T}{\partial t} = \nabla \cdot (\lambda_T\nabla T + \lambda_C\nabla C) - \rho_f c_f\mathbf{v}_r \cdot \nabla T \tag{6.16}$$

where $\rho c = (1-n)\rho_s c_s + n\rho_f c_f$.

6.3   FINITE ELEMENT FORMULATION

6.3.1   Weak forms of fundamental equations

Seepage field

The governing equation of seepage field is assumed to be subjected to the following boundary conditions

Pressure boundary condition

$$p = p_0 \quad \text{on } S_p \tag{6.17}$$

Fluid flux boundary

$$-(n\rho_f \frac{k}{\eta}\nabla p) \cdot \mathbf{n} = \hat{q} \quad \text{on } S_q \tag{6.18}$$

Taking a variation on pressure field δp and integrating by parts, one gets the weak form of Eq. (6.12) as

$$n \left(\int_V \delta p \frac{n}{K_f} \frac{\partial p}{\partial t} dV - \int_V \delta p \beta_{fT} \frac{\partial T}{\partial t} dV - \int_V \delta p \beta_{fC} \frac{\partial C}{\partial t} dV \right)$$

$$+ \int_V \nabla(\delta p) \cdot \frac{k}{\eta} \nabla p dV = \int_{S_q} \delta p \hat{q} dS \tag{6.19}$$

Diffusion field

The diffusion equation is assumed to be subjected to the following boundary conditions

Concentration boundary

$$C = C_0 \quad \text{on } S_C \tag{6.20}$$

Concentration flux boundary

$$-(D_C \nabla C + D_{CT} \nabla T) \cdot \mathbf{n} = \hat{f} \quad \text{on } S_f \tag{6.21}$$

Taking a variation δC and applying the integration by parts to Eq. (6.13), we have the weak form of Eq. (6.13) as

$$\int_V \delta C \frac{\partial C}{\partial t} dV + \int_V \nabla \delta C \cdot D_C \nabla C dV + \int_V \nabla \delta C \cdot D_C \nabla T dV$$

$$+ \int_V \delta C v_r \nabla C dV = \int_{S_q} \delta C \hat{f} dS \tag{6.22}$$

Thermal field

Eq. (6.16) is assumed to be subjected to the following boundary conditions

Temperature boundary

$$T = T_0 \quad \text{on } S_T \tag{6.23}$$

Heat flux boundary

$$-(\lambda_T \nabla T + \lambda_{TC} \nabla C) \cdot \mathbf{n} = \hat{h} \quad \text{on } S_h \tag{6.24}$$

Taking a variation δT on temperature field and applying the integration by parts to the first term of Eq. (6.16), one easily gets its weak form as:

$$\int_V \delta T \rho c \frac{\partial T}{\partial t} dV + \int_V \nabla \delta T \cdot \lambda_T \nabla T dV + \int_V \nabla \delta T \cdot \lambda_{TC} \nabla C dV$$

$$+ \int_V \delta T \rho_f c_f \mathbf{v}_r \cdot \nabla T dV = \int_{S_q} \delta T \hat{h} dS \tag{6.25}$$

6.3.2   Discretization of weak forms

6.3.2.1   *Discretization in physical space*

Pressure, concentration and temperature variables are interpolated in a typical finite element with the use of shape functions as given below:

$$p = [N]\{P\}, \quad C = [N]\{\phi\}, \quad T = [N]\{\chi\} \tag{6.26}$$

Inserting these relations into each respective weak form and after some manipulations, one easily gets the following equation system for each field as

Seepage field

$$[M]_{PP}\{\dot{P}\} + [M]_{PC}\{\dot{\phi}\} + [M]_{PT}\{\dot{\chi}\} + [K]_{PP}\{P\} = \{Q\}_P \tag{6.27}$$

where

$$[M]_{PP} = \int_V \frac{n}{K_f} [N]^T [N] dV; \quad [M]_{PC} = -\int_V n\beta_{fC} [N]^T [N] dV;$$

$$[M]_{PT} = -\int_V n\beta_{fT} [N]^T [N] dV; \quad [K]_{PP} = -\int_V \frac{k}{\eta} [B]^T [B] dV;$$

$$\{Q\}_P = \int_{S_q} N^T \hat{q} dS$$

Diffusion field

$$[M]_{CC}\{\dot{\phi}\} + [K]_{CC}\{\phi\} + [K]_{CT}\{\chi\} = \{Q\}_C \tag{6.28}$$

where

$$[M]_{CC} = \int_V [N]^T[N]dV;$$

$$[K]_{CC} = \int_V D_C[B]^T[B]dV - \int_V \frac{k}{\eta}[N]^T([B]\{P\})^T\{B\}dV [K]_{CT} = \int_V D_{CT}[B]^T[B]dV;$$

$$\{Q\}_C = \int_{S_f} N^T \hat{f} dS$$

Thermal field

$$[M]_{TT}\{\dot{\chi}\} + [K]_{TT}\{\chi\} + [K]_{TC}\{\phi\} = \{Q\}_T \tag{6.29}$$

where

$$[M]_{TT} = \int_V \rho c[N]^T[N]dV$$

$$[K]_{TT} = \int_V \lambda_T[B]^T[B]dV - \int_V \rho_f c_f \frac{k}{\eta}[N]^T ([B]\{P\})^T [B]dV$$

$$[K]_{TC} = \int_V \lambda_{TC}[B]^T[B]dV; \quad \{Q\}_T = \int_{S_h} N^T \hat{h} dS$$

Finally the following simultaneous equation system is obtained for whole domain as

$$[M]\{\dot{X}\} + [K]\{X\} = \{Y\} \tag{6.30}$$

where

$$[M] = \begin{bmatrix} [M]_{PP} & [M]_{PC} & [M]_{PT} \\ [0] & [M]_{CC} & [0] \\ [0] & [0] & [M]_{TT} \end{bmatrix}; \quad [K] = \begin{bmatrix} [K]_{PP} & [0] & [0] \\ [0] & [K]_{CC} & [K]_{CT} \\ [0] & [K]_{TC} & [K]_{TT} \end{bmatrix}$$

$$\{\dot{X}\} = \begin{Bmatrix} \dot{P} \\ \dot{\phi} \\ \dot{\chi} \end{Bmatrix}; \quad \{X\} = \begin{Bmatrix} P \\ \phi \\ \chi \end{Bmatrix}; \quad \{Y\} = \begin{Bmatrix} \{Q\}_P \\ \{Q\}_C \\ \{Q\}_T \end{Bmatrix}$$

6.3.2.2 Discretization in time domain

Although there are several techniques for discretization in time domain, θ-method is chosen herein. Since Eq. (6.30) holds at any time, one may easily write the following relation for a time step $(m + \theta)$ as

$$[M]\{\dot{X}\}_{(m+\theta)} + [K]\{X\}_{(m+\theta)} = \{Y\}_{(m+\theta)} \tag{6.31}$$

With the use of Taylor expansions of variables and after some manipulations as described in Chapter 4, Eq. (6.31) takes the following form:

$$[C^*]\{X\}_{m+1} = \{Y^*\}_{m+1} \tag{6.32}$$

where

$$[C^*] = \left[\frac{1}{\Delta t}[C] + \theta[K] \right];$$

$$\{Y^*\}_{m+1} = \left[\frac{1}{\Delta t}[C] - (1 - \theta)[K] \right] \{X\}_m + \theta\{Y\}_{m+1} + (1 - \theta)\{Y\}_m$$

It should be noted that matrices $[K]_{CC}$ and $[K]_{TT}$ contains unknown variable vector $\{P\}$. Therefore, the resulting equation system is non-linear. However, if time step is sufficiently small, it can be linearized with the use of variable $\{P\}$ of the previous time step.

6.4 EXAMPLES AND DISCUSSIONS

The first example is concerned with the simulation of diffusion-seepage tests on a sandstone sample carried out by Igarashi & Tanaka (1998) under isothermal condition. The pressure gradient in the sample was set to 260 MPa/m. Figure 6.2 shows pressure distribution through the sample at different time steps. Since the pressure distribution attains the steady state in a short period of time, this implies that the diffusion takes

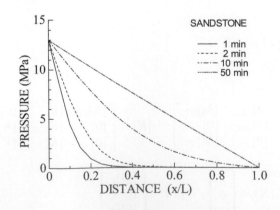

Figure 6.2 Pressure distributions in a sandstone sample.

place under constant fluid velocity field through the sample. Material properties used in the analyses are given in Table 6.1. The computations were carried out for two different situations, specifically,

1) Advection + Conduction and
2) Conduction only.

Figure 6.3 compares the breakdown curves for computed two situations with the experimental one. In the analyses, the effect of diffusion on seepage field is neglected. As seen from the figure, the best fit to the experimental response was obtained from the computation for the advection + conduction situation while much longer period of time is required for conductive diffusion. This fact simply implies that if the diffusion coefficient, which is obtained from a conductive diffusion model under different pressure gradients, will differ from each other.

The method presented in the previous sections is applied to analyze the temperature distribution measured at Armutlu hot springs in the NW Turkey. Pfister *et al.* (1997) were carried out some measurements. They used some models based on the steady state equation of heat transport with advective upward flow to interpret the temperature measurements. For the interpretation of the temperature distribution, an upward velocity of 0.5 m/year was assumed at a depth of 150 m. However, the general temperature profiles of rock mass and nearby thermal spring channels may be different

Table 6.1 Material properties used in analyses.

Fick's diffusion coefficient of solid (m²/day)	5.1×10^{-5}
Fick's diffusion coefficient of fluid (m²/day)	2.1×10^{-4}
Porosity (%)	16
Permeability (m²)	3.1×10^{-12}
Sample length (mm)	50
Sample diameter (mm)	50

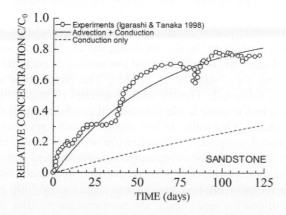

Figure 6.3 Comparison of computed breakdown curves with the experimental curves.

Table 6.2 Properties of materials used in numerical analyses.

Parameter	Schists	Diabase	Air
Thermal conductivity of solid (m^{1}K^{-1})	2.7	0.54	–
Thermal conductivity of fluid (Wm^{-1}K^{-1})	0.6	0.6	0.026
Density of solid (kgm^{-3})	2600	2600	–
Density of fluid (kgm^{-3})	1000	1000	1.2
Specific heat of solid (kJkg^{-1}K^{-1})	0.89	0.89	–
Specific heat of fluid (kJkg^{-1}K^{-1})	4.187	4.187	1.0
Porosity (%)	10	10	100
Heat production rate (Wm^{-3})	1.6×10^{-6}	1.6×10^{-6}	–

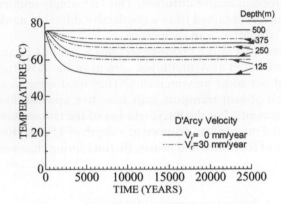

Figure 6.4 Variation of temperatures of some selected points with time.

since the upward fluid velocity could vary due to the permeability rock masses. Three different situations are considered, specifically:

CASE 1: no advection
CASE 2: advection with an upward fluid velocity of 30 mm/year, and
CASE 3: advection with an upward fluid velocity of 100 mm/year.

CASE 1 corresponds to purely conductive heat transport while CASE 2 and CASE 3 to the advective heat transport. In the analyses, an air element next to the ground surface is introduced since the temperature of the surface of rock may not be the same as that of surrounding air (Aydan *et al.*, 1985). The parameters used in numerical analyses are given in Table 6.2. The thermal properties of micaschists and calcerous schist are assumed to be the same while the thermal conductivity of diabase was taken 0.2 times that of schists. This value is less than the value reported by Pfister *et al*. (1997). Since the initial temperature distribution was unknown, the whole domain was initially set to 76°C and the computations were carried out for 25000 years. Figure 6.4 shows the time history of the temperatures of some selected points. As noted from the figure, the temperature becomes almost steady after 7000 years. To compare temperature distributions for three cases, the temperature profiles at 25000 years are selected as the steady state temperature of the vicinity of the well at Armutlu. The

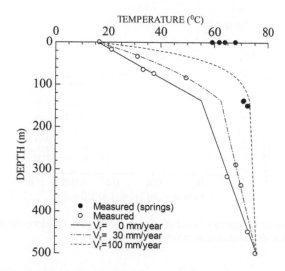

Figure 6.5 Comparison of computed temperature profiles with measured temperatures.

temperature profile for CASE 1 closely predicts the lower bound values of measured temperature profile. On the other hand, the temperature profile for CASE 2 predicts the upper bound values of the measured profile. The actual situation in the vicinity of the Armutlu hot spring may be just in between two predictions (Figure 6.5). The temperature profile for CASE 3 is very close to the temperature profile of hot water in the well, although the profile near the ground surface is a bit different from the actual measurements. Nevertheless, such a situation may be simulated if the rock material is replaced with steel and water in order to simulate the effect of the metal casing of the drill-hole. Furthermore, it should be noted that the upward fluid velocity should be much less than that reported by Pfister *et al.* (1997).

The next example is concerned with moisture transport through a buffer material for radioactive waste disposal under different ambient temperatures, reported by Kanno *et al.* (1999). Recently Basha & Selvadurai (1998) theoretically solved this problem as a spherically symmetric problem.

Kanno *et al.* (1999) used bentonite as a buffer material in their tests and the ambient temperature was varied from 25° to 60°C. They determined Fick's diffusion coefficient of bentonite for moisture transport using a conductive diffusion model. Figure 6.6 shows a simulation of moisture concentration distributions through the sample at different time steps, carried out at an ambient temperature of 25°C. They found that the diffusion coefficients obtained for different ambient temperatures differ from each other.

With the experimental finding of Kanno *et al.* (1999) in mind, a series of parametric studies were carried out. Specifically the following four cases were considered in parametric studies:

CASE 1: Dufour coefficient and Soret coefficient are nil (thermal and diffusion fields are uncoupled)

CASE 2: Dufour coefficient is not nil, but Soret coefficient is nil

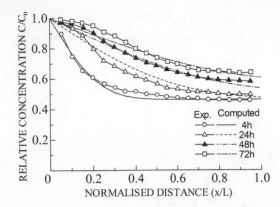

Figure 6.6 Comparison of computed moisture concentration distributions through the sample with experimental distributions at different time steps.

Table 6.3 Material properties used in numerical analyses.

Parameter	CASE 1	CASE 4
Solid's thermal conductivity (Wm^{-1}K^{-1})	0.277	0.277
Fluid's thermal conductivity (Wm^{-1}K^{-1})	0.600	0.600
Dufour's coefficient of solid (Wcm^{-1}kg^{-1})	0	72
Dufour's coefficient of fluid (Wcm^{-1}kg^{-1})	0	72
Soret's coefficient of solid (cm^{2}kg s^{-1}K^{-1})	0	3.6×10^{-5}
Soret's coefficient of solid (cm^{2}kg s^{-1}K^{-1})	0	3.6×10^{-5}
Solid's density (kg m^{-3})	1600	1600
Fluid's density (kg m^{-3})	1000	1000
Solid's specific heat (kJ kg^{-1}K^{-1})	5.0	5.0
Fluid's specific heat (kJ kg^{-1}K^{-1})	1.0	1.0
Fick's diffusion coefficient of solid (cm^2/s)	6.1×10^{-3}	6.1×10^{-3}
Porosity (%)	46.3	46.3

CASE 3: Dufour coefficient is nil, but Soret coefficient is not nil
CASE 4: Dufour coefficient and Soret coefficient are not nil (fully coupled)

Material properties used in analyses are given in Table 6.3 and computed results are shown in Figures 6.7–6.8 for CASE 1 and CASE 4 only. For all cases, temperature distributions through the sample become uniform and the effect of diffusion field on temperature distributions are not observed. CASE 4 is probably the closest situation to actual situations. Particularly, the effects of Duffour coefficient and Soret coefficient are quite remarkable. However, there are almost no experimental results reported on these coefficients to confirm the findings from computations. Therefore, it may be stated that re-assessment of existing experiments with the consideration of Duffour coefficient and Soret coefficient is urgently necessary.

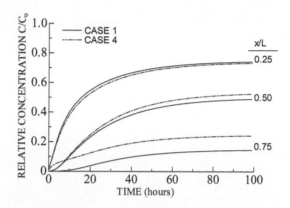

Figure 6.7   Comparison of computed relative concentration versus time curves for CASE 1 and CASE 4.

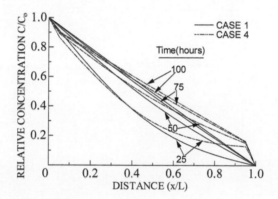

Figure 6.8   Comparison of computed relative concentration distributions at different time steps for CASE 1 and CASE 4.

6.5   CONCLUDING REMARKS

In this chapter, a new mechanical model for fully coupled thermo-hydro-diffusion phenomena in geo-science and geo-engineering on the basis of mixture theory and its finite element representation are presented. The validity of this model is checked through some experiments, which correspond to its special forms. From these comparisons, some shortcomings of determination of Fickian diffusion coefficients from conductive diffusion models are pointed out. Parametric studies on the effect of Duffour and Soret coefficients which are generally neglected in diffusion models indicated that these coefficients must be taken into account in numerical analyses. Such considerations can explain the dependency of Fickian diffusion coefficient on ambient temperature and fluid velocity. Furthermore, there is an urgent experimental necessity to obtain the actual values of Duffour and Soret coefficients for a meaningful assessment of fully coupled thermo-hydro-diffusion phenomena.

Chapter 7

Thermo-hydro-mechanical behaviour of rocks

7.1 INTRODUCTION

As mentioned in Chapter 6, the coupled problems are commonly encountered in geo-science and geo-engineering and the safe disposal of nuclear wastes, petroleum and geothermal exploitations are some of examples for such problems in geo-engineering. Similar problems may also be encountered in earthquake prediction studies in geo-science. The radioactive nuclear waste disposal particularly receives great attention in the field of geoengineering. Rock masses in nature contain pores within the intact rock and discontinuities with apertures of different sizes. One of the approaches to deal with such rock masses is to treat them as multiphase materials.

The theory of elastic wave propagation in a fluid-saturated porous solid was first developed by Biot in 1956 after his study on three dimensional consolidation (1941) in which he had established the equations governing the interaction of the solid and fluid media for quasi-static phenomena. Although Biot's wave propagation theory was based on intuitive ideas and phenomenological concepts, it is still essentially valid and widely used.

Truesdell & Toupin (1960) had introduced the theory of mixtures, which also has been elaborated by many others (i.e. Müller, 1968; Eringen & Ingram, 1965). New theories on the subject have been developed by various researchers besides those based on the theory of mixtures (i.e. Aydan *et al.*, 1995, 1996; Jones, 1975; Bakhvalov & Panasenko, 1984). The theory of mixtures describes the behaviour of multiphase media by considering the effects of interaction between the constituents that make up the whole system. Finite deformation and constitutive relations have been formulated by several authors (i.e. Green & Naghdi, 1967; Morland, 1971; Garg, 1971; Prevost, 1979). Garg & Nur (1973) formulated the constitutive relations for the saturated-porous solid with the mixture theory principles.

In the first part of this chapter, the formulation of coupled thermo-hydro-mechanical behaviour of rock masses is presented using the mixture theory as illustrated in Figure 7.1. Then some practical applications are presented in the rest of the chapter.

7.2 MECHANICAL MODELING BASED ON MIXTURE THEORY

The basics of the mixture theorem lies on the idea of describing the behaviour of the mixture as a whole by considering each single constitute as a continuum for which

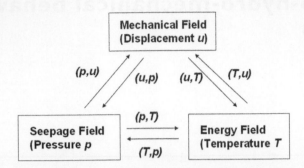

Figure 7.1 Coupling concept of thermo-hydro-mechanics behaviour of rocks.
[u, p]: stress field acts on seepage field through porosity change (pore deformation)
[p, u]: Seepage field acts on the stress field though volumetric strain of rock skeleton and effective stress concept
[T, p]: Temperature variation effects of seepage field through property variation and thermal expansion of pore fluid (convection)
[p, T]: Seepage field acts as the convection on thermal field
[T, u]: Energy field act on stress field though volumetric strain change
[T, u]: Stress field acts on thermal field through mechanical work

the constitutive relations and the kinematic relations are expressed by taking the interactions between the constituents into account (such as diffusion, dissociation, and chemical reactions). In this process, every constituent keeps its own identity, but they are diffused through the material, such that every region, however small on the macroscopic scale, contains some of each constituent.

Truesdell (1969) states the three main metaphysical principles of the theory as quoted below

1   *All properties of the mixture must be mathematical consequences of properties of the constituents.*
2   *So as to describe the motion of a constituent, we may in imagination isolate it from the rest of the mixture provided we allow properly for the actions of the other constituents upon it.*
3   *The motion of the mixture is governed by the same equations as a single body.*

7.2.1 Preliminaries

The motion of a mixture with 2 constituents is described by 2 equations in Eulerian (spatial) terms as,

$$\mathbf{x} = \Phi^{(\alpha)}(\mathbf{x}^{(\alpha)}, t) \quad (\alpha = 1, 2) \tag{7.1}$$

where the functions $\Phi^{(\alpha)}$ are assumed to be sufficiently smooth.

The mass of constituent $s^{(\alpha)}$ per unit volume of the mixture is called its partial density which is denoted by $\rho^{(\alpha)} = \rho^{(\alpha)}(\mathbf{x}, t)$. The total mass per unit volume of composite $\rho(\mathbf{x}, t)$ is given by

$$\rho = \sum_{\alpha} \rho^{(\alpha)} \tag{7.2}$$

Partial mass densities can be written in terms of effective mass densities as

$$\rho^{(\alpha)} = n^{(\alpha)} \rho^{\alpha} \tag{7.3}$$

where ρ^{α} is the effective mass density for the constitute $s^{(\alpha)}$. The effective density can be thought to be the amount of mass of $s^{(\alpha)}$ in a unit volume occupied by $s^{(\alpha)}$ only. The $n^{(\alpha)}$ term appearing in the equation is a concentration factor representing the fraction of $s^{(\alpha)}$ in the mixture.

The velocity field $\mathbf{v}^{(\alpha)}$ defined in the material description (Lagrangian description) is given by

$$\mathbf{v}^{(\alpha)} = \frac{\partial \mathbf{\Phi}^{(\alpha)}}{\partial t} \tag{7.4}$$

While the spatial (or Eulerian) description of the motion is given by

$$\mathbf{v}^{(\alpha)} = \mathbf{v}^{(\alpha)}(\mathbf{x}, t) \tag{7.5}$$

Total mass flow should be equal to the sum of the individual mass flows. This requirement leads us to introduce the mean or barycentric velocity $\mathbf{v}$ of the mixture in the relation stated below as

$$\rho \mathbf{v} = \sum_{\alpha} \rho^{(\alpha)} \mathbf{v}^{(\alpha)} \tag{7.6}$$

Then, the velocity of the constituent $s^{(\alpha)}$ relative to the mean velocity field can be expressed as

$$\mathbf{w}^{(\alpha)} = \mathbf{v}^{(\alpha)} - \mathbf{v} \tag{7.7}$$

Using the relation (7.6) we can write

$$\sum_{\alpha} \rho^{\alpha} \mathbf{w}^{(\alpha)} = 0 \tag{7.8}$$

The relative velocity $\mathbf{w}^{(\alpha)}$ is called the diffusion velocity of the constituent $s^{(\alpha)}$.

The material time derivatives $D^{(\alpha)}/Dt$, D/Dt are given for arbitrary scalar and vector functions $\psi(\mathbf{x}, t)$ and $\mathbf{u}(\mathbf{x}, t)$ as

$$\frac{D^{(\alpha)}\psi}{Dt} = \frac{\partial \psi}{\partial t} + \mathbf{v}^{(\alpha)} \cdot \operatorname{grad} \psi, \quad \frac{D\psi}{Dt} = \frac{\partial \psi}{\partial t} + \mathbf{v} \cdot \operatorname{grad} \psi \tag{7.9a}$$

$$\frac{D^{(\alpha)}\mathbf{u}}{Dt} = \frac{\partial \mathbf{u}}{\partial t} + (\mathrm{grad}\,\mathbf{u})\mathbf{v}^{(\alpha)}, \quad \frac{D\mathbf{u}}{Dt} = \frac{\partial \mathbf{u}}{\partial t} + (\mathrm{grad}\,\mathbf{u})\mathbf{v} \qquad (7.9b)$$

The derivative $D^{(\alpha)}/Dt$ follows the motion of the constituent $s^{(\alpha)}$ and the derivative D/Dt follows the mean motion of the mixture. For each α, β ($\alpha = 1, 2, \beta = 1, 2$) we can write

$$\frac{D^{(\alpha)}\psi}{Dt} = \frac{D^{(\beta)}\psi}{Dt} + (\mathbf{v}^{(\alpha)} - \mathbf{v}^{(\beta)}) \cdot \mathrm{grad}\,\psi = \frac{D\psi}{Dt} + \mathbf{u} \cdot \mathrm{grad}\,\psi \qquad (7.10a)$$

$$\frac{D^{(\alpha)}\mathbf{w}}{Dt} = \frac{D^{(\beta)}\mathbf{w}}{Dt} + (\mathrm{grad}\,\mathbf{w})(\mathbf{v}^{(\alpha)} - \mathbf{v}^{(\beta)}) = \frac{D\mathbf{w}}{Dt} + (\mathrm{grad}\,\mathbf{w})\mathbf{u}^{(\alpha)} \qquad (7.10b)$$

By using the relations (7.3), (7.10) and the material derivative definitions (7.9), it follows that

$$\sum_{\alpha} \rho^{(\alpha)} \frac{D^{(\alpha)}\psi}{Dt} = \rho \frac{D\psi}{Dt}, \quad \sum_{\alpha} \rho^{(\alpha)} \frac{D^{(\alpha)}\mathbf{u}}{Dt} = \rho \frac{D\mathbf{u}}{Dt} \qquad (7.11)$$

7.2.2   Definitions of thermo-hydro-mechanical quantities for fluid-saturated porous media

On the basis of the principles of mixture theory equations governing the behaviour of the coupled fluid-saturated porous solid system are derived herein. The superscripts (s) and (f) will be employed referring to solid and fluid phases respectively, which correspond to $\alpha = 1$ and $\alpha = 2$ in the general mixture theorem expressions.

(a) Partial and total densities and mass fluxes

According to (7.2), the total mass density of the bulk material in terms of the partial mass densities of solid and fluid will be

$$\rho = \rho^{(s)} + \rho^{(f)} \qquad (7.12)$$

Denoting the material porosity by n, by the relation (7.2), the partial densities of solid and fluid in terms of the material mass densities ρ^s and ρ^f can be expressed as

$$\rho^{(s)} = (1 - n)\rho^s \qquad (7.13a)$$

$$\rho^{(f)} = n\rho^f \qquad (7.13b)$$

Total and partial mass fluxes are related to each other as follow

$$\mathbf{q} = \mathbf{q}^{(s)} + \mathbf{q}^{(f)} \qquad (7.14)$$

where

$$\mathbf{q}^{(s)} = (1 - n)\rho^s \mathbf{v}^s \quad \text{and} \quad \mathbf{q}^{(f)} = n\rho^f \mathbf{v}^f.$$

(b) Partial and total stress tensors

The total stress σ acting on the continuum region under consideration can be regarded as the summation of the partial stresses

$$\sigma = \sum_{\alpha} \sigma^{(\alpha)} \tag{7.15}$$

where $\sigma^{(\alpha)}$ can be expressed in terms of the effective stress σ^α and the area fraction $n^{(\alpha)}$ as

$$\sigma^{(\alpha)} = n^{(\alpha)} \sigma^\alpha \tag{7.16}$$

Therefore the total stress σ acting on the two phase bulk material can be expressed in terms of partial effective stress acting on the solid skeleton and on the fluid

$$\sigma = \sigma^{(s)} + \sigma^{(f)} = (1 - n)\sigma^s + n\sigma^f \tag{7.17}$$

Let us denote the partial effective stress on the solid skeleton as given below

$$\sigma'_{ij} = (1 - n)(\sigma^s_{ij} + p\delta_{ij}) \tag{7.18}$$

and let us denote the effective partial stress on the fluid as

$$\sigma^f = -\mathbf{p}, \quad p_{ij} = -p\delta_{ij} \tag{7.19}$$

The negative sign is due to the fluid pressure being always compressive. Then equation (7.17) can be written in the form given below as

$$\sigma_{ij} = (1 - n)\sigma^s_{ij} - np\delta_{ij} = \sigma'_{ij} - p\delta_{ij} \quad \text{or} \quad \sigma = \sigma' - \mathbf{p} \tag{7.20}$$

where p is interpreted as the total pore fluid pressure acting on the fluid parts of a face on any section of the bulk material.

(c) Heat

Total temperature and heat flux vector of the bulk material are given in the following form (see Sub-section 6.2.1)

$$T = (1 - n)T^s + nT^f \tag{7.21}$$

$$\mathbf{h} = (1 - n)\mathbf{h}^s + n\mathbf{h}^f \tag{7.22}$$

Similarly, total energy production is related to those of the solid and fluid phases as follow

$$Q = (1 - n)Q^s + nQ^f \tag{7.23}$$

7.2.3   Mass conservation law for two-phase materials

The mass of a constituent is not necessarily conserved due to the occurrence of chemical reactions between the constituents. Then the balance of mass for the constituent $s^{(\alpha)}$ is postulated as

$$\frac{d}{dt}\int_R \rho^{(\alpha)}dV + \int_{\partial R} \rho^{(\alpha)}\mathbf{v}^{(\alpha)} \cdot \mathbf{n}dA = \int_R m^{(\alpha)}dV \qquad (7.24)$$

where $m^{(\alpha)}$ is the density of mass production for the constituent $s^{(\alpha)}$ arising from all the other constituents. The unit outward normal to the surface ∂R is denoted by $\mathbf{n}$. It is assumed that, for the mixture as a whole the total mass is conserved. Then the balance of mass for the mixture can be written as

$$\frac{d}{dt}\int_R \sum_{\alpha} \rho^{(\alpha)}dV + \int_{\partial R} \sum_{\alpha} \rho^{(\alpha)}\mathbf{v}^{(\alpha)} \cdot \mathbf{n}dA = 0 \qquad (7.25)$$

Note that summation of the density of mass production of all the constituents in the mixture should satisfy

$$\sum_{\alpha} m^{(\alpha)} = 0 \qquad (7.26)$$

Assuming no chemical reactions occurring between the solid and the fluid, the balance of mass for the solid phase and the fluid phase can be written as

$$\frac{d}{dt}\int_R \rho^{(s)}dV + \int_{\partial R} \rho^{(s)}\mathbf{v}^{(s)} \cdot \mathbf{n}dA = 0 \qquad (7.27a)$$

$$\frac{d}{dt}\int_R \rho^{(f)}dV + \int_{\partial R} \rho^{(f)}\mathbf{v}^{(f)} \cdot \mathbf{n}dA = 0 \qquad (7.27b)$$

On using the Gauss' divergence theorem and converting the surface integrals into volume integrals and assuming the integrands are continuous on the region R in the Eq. (7.27), we can write

$$\frac{\partial \rho^{(s)}}{\partial t} + \mathbf{v}^{(s)} \cdot \nabla \rho^{(s)} + \rho^{(s)}\nabla \cdot \mathbf{v}^{(s)} = 0 \qquad (7.28a)$$

$$\frac{\partial \rho^{(f)}}{\partial t} + \mathbf{v}^{(f)} \cdot \nabla \rho^{(f)} + \rho^{(f)}\nabla \cdot \mathbf{v}^{(f)} = 0 \qquad (7.28b)$$

Then, the balance of mass expression for the bulk material is written as

$$\frac{D^{(s)}\rho^{(s)}}{Dt} + \frac{D^{(f)}\rho^{(f)}}{Dt} + \rho^{(s)}\nabla \cdot \mathbf{v}^{(s)} + \rho^{(f)}\nabla \cdot \mathbf{v}^{(f)} = 0 \qquad (7.29)$$

More specifically, Equation (7.29) can be re-written after some manipulations as

$$\frac{1-n}{\rho^s}\frac{\partial \rho^s}{\partial t} + \frac{n}{\rho^f}\frac{\partial \rho^f}{\partial t} + \frac{1}{\rho^s}\nabla \cdot \mathbf{q}^s + \frac{1}{\rho^f}\nabla \cdot \mathbf{q}^f = 0 \tag{7.30}$$

7.2.4 The equations of momentum balance

The forces acting on the constituent $s^{(\alpha)}$ are

1 Body forces (such as gravity forces)
2 The force exerted by the mixture outside the region R on the constituent $s^{(\alpha)}$.

The vector field $\mathbf{t}^{(\alpha)}(\mathbf{n}, \mathbf{x}, t)$ defined on ∂R which is measured per unit area of ∂R is called the critical stress vector (traction) and accounts for the effect of the mixture outside the region R upon the constituent $s^{(\alpha)}$. In the momentum balance of the constituent $s^{(\alpha)}$, the momentum supplied to $s^{(\alpha)}$ due to chemical reactions with the other constituents and the momentum transfer due to other interaction effects such as the relative motion of the constituents have to be taken into consideration.

According to the above discussion the equation of linear momentum balance for the constituent $s^{(\alpha)}$ can be expressed as

$$\frac{d}{dt}\int_R \rho^{(\alpha)}\mathbf{v}^{(\alpha)}dV + \int_{\partial R}\rho^{(\alpha)}\mathbf{v}^{(\alpha)}(\mathbf{v}^{(\alpha)}\cdot\mathbf{n})dA - \int_R m^{(\alpha)}J^{(\alpha)}dV$$

$$= \int_R(\mathbf{b}^{(\alpha)} + \boldsymbol{\psi}^{(\alpha)})dV + \int_{\partial R}\mathbf{t}^{(\alpha)}dA \tag{7.31}$$

where

$\mathbf{b}^{(\alpha)}$: the external body force acting on $s^{(\alpha)}$ measured per unit mass of $s^{(\alpha)}$.
$J^{(\alpha)}$: accounts for the effect of the momentum supplied to $s^{(\alpha)}$ due to chemical reactions and has the dimension of velocity
$\boldsymbol{\psi}^{(\alpha)}$: diffusive force exerted on $s^{(\alpha)}$ by the other constituents

Considering that the overall mass is conserved for the mixture and regarding the force $\boldsymbol{\psi}^{(\alpha)}$ as an internal effect, the balance of linear momentum for the mixture can be expressed as

$$\frac{d}{dt}\int_R\sum_\alpha\rho^{(\alpha)}\mathbf{v}^{(\alpha)}dV + \int_{\partial R}\sum_\alpha\rho^{(\alpha)}\mathbf{v}^{(\alpha)}(\mathbf{v}^{(\alpha)}\cdot\mathbf{n})dA$$

$$= \int_R\sum_\alpha\mathbf{b}^{(\alpha)}dV + \int_{\partial R}\sum_\alpha\mathbf{t}^{(\alpha)}dA \tag{7.32}$$

Now let us consider the 2 phase mixture of fluid-saturated porous solid with the same assumptions and considerations used in the expression of mass balance. The balance of linear momentum for the solid phase and the fluid phase can be written as

$$\frac{d}{dt}\int_R\rho^{(s)}\mathbf{v}^{(s)}dV + \int_{\partial R}\rho^{(s)}\mathbf{v}^{(s)}(\mathbf{v}^{(s)}\cdot\mathbf{n})dA = \int_R(\boldsymbol{\psi}^{(s)} + \mathbf{b}^{(s)})dV + \int_{\partial R}\mathbf{t}^{(s)}dA \tag{7.33a}$$

$$\frac{d}{dt}\int_R \rho^{(f)}\mathbf{v}^{(f)}dV + \int_{\partial R}\rho^{(f)}\mathbf{v}^{(f)}(\mathbf{v}^{(f)}\cdot\mathbf{n})dA = \int_R(\boldsymbol{\psi}^{(f)}+\mathbf{b}^{(f)})dV + \int_{\partial R}\mathbf{t}^{(f)}dA \quad (7.33b)$$

On using Gauss' divergence theorem and assuming the integrands are continuous on region R, we can write (7.33) as

$$\rho^{(s)}\left(\frac{\partial\mathbf{v}^{(s)}}{\partial t}+\mathbf{v}^{(s)}\cdot\nabla\mathbf{v}^{(s)}\right)=\boldsymbol{\psi}^{(s)}+\nabla\cdot\boldsymbol{\sigma}^{(s)}+\mathbf{b}^{(s)} \quad (7.34a)$$

$$\rho^{(f)}\left(\frac{\partial\mathbf{v}^{(f)}}{\partial t}+\mathbf{v}^{(f)}\cdot\nabla\mathbf{v}^{(f)}\right)=\boldsymbol{\psi}^{(f)}+\nabla\cdot\boldsymbol{\sigma}^{(f)}+\mathbf{b}^{(f)} \quad (7.34b)$$

where $\mathbf{t}^{(s)}=\boldsymbol{\sigma}^{(s)}\cdot\mathbf{n}$ and $\mathbf{t}^{(f)}=\boldsymbol{\sigma}^{(f)}\cdot\mathbf{n}$.

With $\boldsymbol{\psi}^{(s)}=-\boldsymbol{\psi}^{(f)}$, balance of linear momentum for the bulk material, then can be written as

$$(1-n)\rho^s\frac{d\mathbf{v}^{(s)}}{dt}+n\rho^f\frac{d\mathbf{v}^{(f)}}{dt}==\nabla\cdot\boldsymbol{\sigma}+\mathbf{b} \quad (7.35)$$

where

$$\frac{d\mathbf{v}^{(s)}}{dt}=\frac{\partial\mathbf{v}^s}{\partial t}+\mathbf{v}^s\cdot\nabla\mathbf{v}^s \quad \frac{d\mathbf{v}^{(f)}}{dt}=\frac{\partial\mathbf{v}^f}{\partial t}+\mathbf{v}^f\cdot\nabla\mathbf{v}^f$$

$$\boldsymbol{\sigma}=\boldsymbol{\sigma}^{(s)}+\boldsymbol{\sigma}^{(f)}=\boldsymbol{\sigma}'-\mathbf{p}; \quad \mathbf{b}=\mathbf{b}^{(s)}+\mathbf{b}^{(f)}=(1-n)\mathbf{b}^s+n\mathbf{b}^f$$

For fluid phase, the final expression is

$$\nabla\cdot\boldsymbol{\sigma}^{(f)}+n\mathbf{b}^f=n\rho^f\dot{\mathbf{v}}^{(f)}+\frac{\eta}{k}\mathbf{w} \quad \text{or} \quad -\nabla\cdot(np)+n\mathbf{b}^f=n\rho^f\dot{\mathbf{v}}^{(f)}+\frac{\eta}{k}\mathbf{w} \quad (7.36)$$

where $\mathbf{w}=\mathbf{v}^f-\mathbf{v}^s$. As it can be noticed easily, the equations of motion we have derived for the bulk material as a whole and for the pore fluid are the same as the equations of motion proposed by Biot (1942), intuitively.

7.2.5 Energy conservation law

Let us consider the 2 phase mixture of fluid-saturated porous solid with the same assumptions and considerations used in the expression of energy conservation. The conservation of energy for the solid phase and the fluid phase can be written as

Solid phase

$$\frac{d}{dt}\int_R\rho^{(s)}U^{(s)}dV=-\int_R\nabla\cdot\mathbf{h}^{(s)}dV+\int_R\boldsymbol{\sigma}^{(s)}:(\nabla\mathbf{v}^s)dV+\int_R Q^{(s)}dV+\int_R\eta_{sf}dV \quad (7.37a)$$

Fluid phase

$$\frac{d}{dt}\int_R\rho^{(f)}U^{(f)}dV=-\int_R\nabla\cdot\mathbf{h}^{(f)}dV+\int_R\boldsymbol{\sigma}^{(f)}:(\nabla\mathbf{v}^f)dV+\int_R Q^{(f)}dV+\int_R\eta_{fs}dV \quad (7.37b)$$

Summing Eqs. (7.37a) and (7.37b) together with $\eta_{sf} = -\eta_{fs}$, the energy conservation law for the bulk material takes the following form

$$(1-n)\,\rho^s\frac{dU^s}{dt} + n\rho^f\frac{dU^f}{dt} = \nabla \cdot \mathbf{h} + (1-n)\boldsymbol{\sigma}^s:\nabla\mathbf{v}^s + n\boldsymbol{\sigma}^f:\nabla\mathbf{v}^f + Q \qquad (7.38)$$

where $\mathbf{h} = (1-n)\mathbf{h}^s + n\mathbf{h}^f$ and $Q = (1-n)\,Q^s + nQ^f$.

7.2.6 Constitutive laws

(a) Mechanical field

The constitutive law between effective stress and effective strain tensors are given in the following form using Hooke's law

$$\boldsymbol{\sigma}' = \mathbf{D} : \boldsymbol{\varepsilon}' \qquad (7.39)$$

where

$$\boldsymbol{\varepsilon}' = \boldsymbol{\varepsilon} - \boldsymbol{\varepsilon}_P - \boldsymbol{\varepsilon}_T - \boldsymbol{\varepsilon}_o$$

$$\boldsymbol{\varepsilon} = \frac{1}{2}\left(\nabla\mathbf{u} + (\nabla\mathbf{u})^T\right),$$

$\mathbf{D}$ is elasticity tensor,
$\boldsymbol{\varepsilon}_P$ is volumetric strain due pore pressure,
$\boldsymbol{\varepsilon}_T$ is volumetric thermal strain due to temperature variation,
$\boldsymbol{\varepsilon}_o$ is volumetric strain due to chemical actions or electrical attractions.

They are given specifically as follow:

$$\boldsymbol{\varepsilon}_P = -\frac{1}{3K_s}p\mathbf{I}, \quad \boldsymbol{\varepsilon}_T = \frac{\beta_s}{3}(T - T_0)\mathbf{I} \qquad (7.40)$$

where K_s and β_s are bulk stiffness and thermal expansion coefficients. The strain due to chemical reaction/electrical attraction would be omitted in this chapter. Nevertheless, it could be envisaged similar to that described in Chapter 3.

(b) Seepage field

Due to the relative motion of the fluid with respect to the solid skeleton, the fluid flow is considered to undergo a Darcy type of resistance. The relative flow of the fluid with respect to solid in terms of volume per unit area of the bulk medium is expressed as

$$\mathbf{w} = n(\mathbf{v}^f - \mathbf{v}^s) \qquad (7.41)$$

Darcy law is usually used to relate the relative average fluid velocity to pressure in the following form

$$\mathbf{w} = n(\mathbf{v}^f - \mathbf{v}^s) = -\frac{k}{\rho_o^f g}(\nabla p + \rho^f g \nabla \varsigma) \qquad (7.42)$$

where k is hydraulic conductivity and ρ_o^f is initial density, g is gravitational acceleration and $\varsigma = \varsigma_o - z$.

(c) Thermal field

Fourier's law is well known as a constitutive law for associating heat flux to temperature gradient as used in Chapter 6. This law is also expanded by associating heat flux $\mathbf{h_s}$ with temperature gradient as

$$\mathbf{h}^s = -\lambda^s \nabla T^s \qquad (7.43)$$

where λ^s is thermal conductivity coefficient of solid phase. Similarly the following relation can be written for fluid phase.

$$\mathbf{h}^f = -\lambda^f \nabla T^f \qquad (7.44)$$

where λ^f is thermal conductivity coefficient of fluid phase. If $T^s = T^f = T$, then average heat flux $\mathbf{h}$ takes the following form.

$$\mathbf{h} = -\lambda_T \nabla T \qquad (7.45)$$

where
$\lambda_T = (1 - n)\lambda^s + n\lambda^f$. λ_T is average conductivity coefficient.

(d) Dependency of density and viscosity of fluid phase to pressure and temperature variations

The variation of density of fluid phase is assumed to be linearly dependent on the variation of temperature and pressure changes as given below (Seiki, 1994; Seiki *et al.*, 1996):

$$\rho^f = \rho_o^f \left(1 - \beta_f (T - T_o) + \frac{1}{K_f}(p - p_o)\right) \qquad (7.46)$$

The viscosity of the fluid phase (i.e. water) depends upon the temperature variations. Huyakorn & Pinder (1983) suggested the following relation for the variation of viscosity of water

$$\eta = 239.4 \times 10^{248.37/(T+131.15)} \times 10^{-6} \qquad (7.47)$$

Unit is (g/cm/sec).

(e) Relation for time derivatives of density of solid and fluid phases

Time derivatives of density of solid and fluid phases in relation to effective stress, fluid pressure and temperature fields are given in the following form:

$$\frac{\partial \rho_s}{\partial t} = -\frac{\rho_s}{3K_s(1-n)}\frac{\partial \bar{\sigma}'}{\partial t} + \frac{\rho_s}{K_s}\frac{\partial p}{\partial t} - \rho_s\beta_s\frac{\partial T}{\partial t} \tag{7.48}$$

$$\frac{\partial \rho^f}{\partial t} = \frac{\rho_0^f}{K_f}\frac{\partial p}{\partial t} - \rho_0^f\beta_f\frac{\partial T}{\partial t} \tag{7.49}$$

$\bar{\sigma}'$ is average effective stress and it is given as

$$\bar{\sigma}' = tr\sigma' \tag{7.50}$$

Furthermore, the time derivative volumetric strain of solid phase is given as

$$\nabla \cdot \mathbf{v}^s = \dot{\varepsilon}^s = tr\dot{\varepsilon} \tag{7.51}$$

7.2.7 Final governing equations

Seepage field

The final form of seepage field can be obtained through the utilization of Eqs. (7.30), and 7.48 to 7.51 as

$$\left\{ tr\dot{\varepsilon} - \frac{1}{3K_s}tr(\mathbf{D}{:}\dot{\varepsilon}) \right\} + \left(\frac{1-n}{K_s} + \frac{n}{K_f} - \frac{1}{9K_s^2}tr(\mathbf{D}{:}\mathbf{I}) \right)\frac{\partial p}{\partial t}$$

$$+ \left\{ -(1-n)\beta_s - n\beta_f + \frac{\beta_s}{9K_s}tr(\mathbf{D}{:}\mathbf{I}) \right\}\frac{\partial T}{\partial t} - \nabla \cdot \left\{ \frac{\eta_0}{\eta}\frac{k}{\rho_0^f g}(\nabla p + \rho^f g \nabla \varsigma) \right\} = 0 \tag{7.52}$$

Mechanical field

Final form of the governing equation for mechanical field takes the following form using Eqs. 7.35, 7.39 and 7.40 and neglecting the inertia components as

$$\nabla \cdot \left(\mathbf{D}{:}\varepsilon + \left(\frac{p}{3K_s} - \frac{\beta_s}{3}(T - T_o) \right)\mathbf{D}{:}\mathbf{I} \right) + \nabla p + (1-n)\mathbf{b}^s + n\mathbf{b}^f = 0 \tag{7.53}$$

Thermal field

Internal energies of solid and fluid phases may be related to temperature field with the use of specific heat coefficients c_s and c_f as defined in Chapter 6

Solid Phase

$$\frac{\partial U_s}{\partial t} = \frac{\partial U_s}{\partial T}\frac{\partial T}{\partial t} = c_s\frac{\partial T}{\partial t} \tag{7.54}$$

Fluid Phase

$$\frac{\partial U_f}{\partial t} = \frac{\partial U_f}{\partial T}\frac{\partial T}{\partial t} = c_f\frac{\partial T}{\partial t} \tag{7.55}$$

With the use of equations above, Eq. (7.43–7.45), Eq. (7.38) becomes

$$\{(1-n)\rho_s C_s + n\rho_f C_f\}\frac{\partial T}{\partial t} + \{(1-n)\rho^s C_s v^s + n\rho^f C_f v^f\}\cdot\nabla T - \nabla\cdot(\lambda\nabla T) = 0 \tag{7.56}$$

The term $(nv^f\cdot\nabla T)$ associated with fluid flow and it may be related to Darcy's law as

$$nv^f\cdot\nabla T \cong n(v^f - v^s)\cdot\nabla T = -\frac{\eta_0}{\eta}\frac{k}{\rho_0^f g}(\nabla p + \rho^f g\nabla\varsigma)\cdot\nabla T \tag{7.57}$$

7.3 FINITE ELEMENT FORMULATION

The formulation would basically follow the same procedures as described in previous Chapters 4, 5 and 6. Therefore, some manipulations are omitted in this section

7.3.1 Weak forms of fundamental equations

(a) Seepage field

The governing equation of seepage field is assumed to be subjected to the following boundary conditions

Pressure boundary condition

$$p = p_0 \quad \text{on } S_p \tag{7.58}$$

Fluid flux boundary

$$-\left(\frac{\eta_0}{\eta}\frac{k}{\rho_0^f g}(\nabla p + \rho^f g\nabla\varsigma)\right)\cdot\mathbf{n} = \hat{q}_H \quad \text{on } S_q \tag{7.59}$$

Taking a variation on pressure field δp and integrating by parts, one gets the weak form of Eq. (7.52) as

$$\int_V \delta p\left\{tr\dot{\varepsilon} - \frac{1}{3K_s}tr(\mathbf{D}:\dot{\varepsilon})\right\}dV + \int_V \delta p\left(\frac{1-n}{K_s} + \frac{n}{K_f} - \frac{1}{9K_s^2}tr(\mathbf{D}:\mathbf{I})\right)\frac{\partial p}{\partial t}dV$$

$$+\int_V \delta p\left\{-(1-n)\,\beta_s - n\beta_f + \frac{\beta_s}{9K_s}tr(\mathbf{D}:\mathbf{I})\right\}\frac{\partial T}{\partial t}dV$$

$$+\int_V \nabla\delta p\cdot\left\{\frac{\eta_0}{\eta}\frac{k}{\rho_f^0 g}(\nabla p + \rho_f g\nabla\varsigma)\right\}dV + \int_{S_q} \delta p\hat{q}_H dS = 0 \tag{7.60}$$

(b) Thermal field

Eq. (7.56) is assumed to be subjected to the following boundary conditions

Temperature boundary

$$T = T_0 \quad \text{on } S_T \tag{7.61}$$

Heat flux boundary

$$-\left(\lambda_T \nabla T + \frac{\eta_0}{\eta} \frac{k}{\rho_0^f g}(\nabla p + \rho^f g \nabla \varsigma) T\right) \cdot \mathbf{n} = \hat{q}_T + \gamma(T - \overline{T}) \quad \text{on } S_h \tag{7.62}$$

Taking a variation δT on temperature field and applying the integration by parts to the first term of Eq. (7.56), one easily gets its weak form as:

$$\int_V \delta T\{(1-n)\rho_s C_s + n\rho_f C_f\} \frac{\partial T}{\partial t} dV + \int_V \nabla \delta T \cdot (\lambda \nabla T) \, dV$$

$$+ \int_V \nabla \delta T \cdot \left\{\rho^f C_f \frac{\eta_0}{\eta} \frac{k}{\rho_0^f g}(\nabla p + \rho^f g \nabla \varsigma)\right\} T dV + \int_{S_s} \delta T \hat{q}_T dS$$

$$+ \int_{S_m} \delta T \gamma (T - \overline{T}) dS = 0 \tag{7.63}$$

(c) Mechanical field

Eq. (7.53) is assumed to be subjected to the following boundary conditions

Displacement boundary

$$\dot{\mathbf{u}} = \dot{\mathbf{u}}_0 \quad \text{on } S_p \tag{7.64}$$

Traction Boundary

$$(\dot{\sigma}' - \dot{\mathbf{p}}) \cdot \mathbf{n} = \dot{\hat{\mathbf{t}}} \quad \text{on } S_t \tag{7.65}$$

Taking a variation $\delta \mathbf{u}$ on displacement field and applying the integration by parts to the first term of Eq. (7.53), one easily gets its weak form as:

$$\int_{S_t} \delta \mathbf{u} \cdot \dot{\hat{\mathbf{t}}} dS - \int_V \delta \varepsilon \cdot \left(\mathbf{D}:\dot{\varepsilon} + \left(\frac{\dot{p}}{3K_s} - \frac{\beta_s}{3}\dot{T}\right)\mathbf{D}:\mathbf{I} - \dot{p}\mathbf{I}\right) dV$$

$$+ \int_V \delta \mathbf{u} \cdot n\rho_o^f \mathbf{g}\left(-\beta_f \dot{T} + \frac{\dot{p}}{K_s}\right) dV = 0 \tag{7.66}$$

7.3.2 Discretization of weak forms

7.3.2.1 *Discretization in physical space*

Pressure, displacement and temperature variables are interpolated in a typical finite element with the use of shape functions. It should be noted that the order of shape

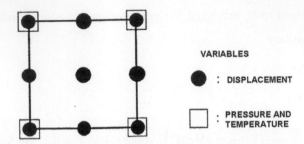

Figure 7.2 Illustration of number of nodes of elements associated with physical discretization.

functions would be different for displacement, temperature and pressure variables. The shape function of displacement field should be one order higher as compared with those for temperature and pressure field (See Appendix for a specific example and discussions). Figure 7.2 illustrates such concept for nine (N) and four-noded (M) isoparametric elements.

Inserting these relations into each respective weak form and after some manipulations, one easily gets the following equation system for each field as

$$[C]_{PU}\{\dot{U}\} + [C]_{PP}\{\dot{P}\} + [C]_{PT}\{\dot{T}\} + [K]_{PP}\{P\} = \{Q\} \tag{7.67a}$$

$$[C]_{UU}\{U\} + [C]_{UT}\{T\} + [C]_{UP}\{P\} = \{\dot{F}\} \tag{7.67b}$$

$$[C]_{TT}\{\dot{T}\} + [K]_{TT}\{T\} = \{S\} \tag{7.67c}$$

where

$$[C]_{PP} = \left(\frac{1-n}{K_s} + \frac{n}{K_f}\right)\int_V [M]^T[M]dV + \int_V \frac{1}{9K_s^2}[M]^T[L]_1^T[D][L]_1[M]dV$$

$$[C]_{PU} = \int_V \left\{\frac{1}{3K_s}[M]^T[L]_1^T[D][B] - [M]^T[L]_1^T[B]\right\}dV$$

$$[C]_{PT} = \int_V ((1-n)\beta_s - n\beta_f)[M]^T[M]dV - \int_V \frac{\beta_s}{9K_s}[M]^T[L]_1^T[D][L]_1[M]dV$$

$$[K]_{PP} = \int_V \frac{\eta_0}{\eta}\frac{1}{\rho_o^f g}[A]^T([M]\{\eta\}^{-1})\{k\}[A]dV$$

$$\{Q\} == \int_{S_q} [M]^T\hat{q}_H dS + \int_V \frac{\eta_0}{\rho_o^f g}[A]^T([M]\{\eta\}^{-1})\{\eta\}([M]\{\rho^f\})g[A]\{\zeta\}dV$$

$$[C]_{UP} = \int_V \left\{\frac{1}{3K_s}[B]^T[D][L]_1^T[M] - [B]^T[L]_1^T[M]\right\}dV - \int_V n\frac{\rho_o^f g}{K_f}[N]^T[L]_2[M]dV$$

$$[C]_{UU} = \int_V [B]^T [D][B] dV$$

$$[C]_{UT} = \int_V \left\{ \frac{\beta_s}{3} [B]^T [D][L]_1^T [M] \right\} dV - \int_V n\rho_o^f g\beta_f [N]^T [L]_2 [M] dV$$

$$\{\dot{F}\} = \int_{S_t} [N]^T \{\dot{t}\} dS$$

$$[K]_{TT} = \int_V [A]^T ([M]\{\rho^f\} C_f \eta_o ([M]\{\eta\}^{-1}) \frac{k}{\rho_o^f g} ([A]\{P\} + ([M]\{\rho^f\})g[A]\{\zeta\})[M] dV$$
$$+ \int_V \lambda [A]^T [A] dV + \int_{S_h} \gamma [M]^T [M] dV$$

$$S = -\int_{S_S} [M]^T \{\hat{q}_T\} dS + \int_{S_h} \gamma [M]^T \{\overline{T}\} dS$$

$$[A] = \nabla [M]; \quad [B] = \nabla [N]; \quad [L]_1^T = [1 \quad 1 \quad 1]; \quad [L]_2^T = [0 \quad 0 \quad -1]$$

$\{P\}, \{U\}, \{T\}, \{\varsigma\}, \{\rho\}^f$ and $\{\eta\}$ are vectors of variables of nodal pressure, displacement, temperature, height from reference plane and density of fluid.

The relation above can be re-written in a compact form as follows

$$\begin{bmatrix} C_{PP} & C_{PU} & C_{PT} \\ C_{UP} & C_{UU} & C_{UT} \\ 0 & 0 & C_{TT} \end{bmatrix} \begin{Bmatrix} \dot{P} \\ \dot{U} \\ \dot{T} \end{Bmatrix} + \begin{bmatrix} K_{PP} & 0 & 0 \\ 0 & 0 & 0 \\ 0 & 0 & K_{TT} \end{bmatrix} \begin{Bmatrix} P \\ U \\ T \end{Bmatrix} = \begin{Bmatrix} Q \\ \dot{F} \\ S \end{Bmatrix} \tag{7.68a}$$

or

$$[C]\{\dot{X}\} + [K]\{X\} = \{Y\} \tag{7.68b}$$

7.3.2.2 Discretization in time domain

Similar to the approaches described in Chapters 4 to 6, if θ-method is chosen for discretization in time domain Eq. (7.68) takes the following form:

$$[H^*]\{X\}_{m+1} = \{Y^*\}_{m+1} \tag{7.69}$$

where

$$[H^*] = \left[\frac{1}{\Delta t} [C] + \theta [K] \right];$$

$$\{Y^*\}_{m+1} = \left[\frac{1}{\Delta t} [C] - (1-\theta)[K] \right] \{X\}_m + \theta \{Y\}_{m+1} + (1-\theta) \{Y\}_m$$

It should be noted that matrices $[K]_{TT}$ contains unknown variable vector $\{P\}$. Therefore, the resulting equation system is non-linear. However, if time step is sufficiently small, it can be linearized with the use of variable $\{P\}$ of the previous time step.

7.4 EXAMPLES AND DISCUSSIONS

7.4.1 Example of buried heat source in fully saturated shallow rock mass

The first application of the theory and its numerical method presented in the previous sections involves the displacement, porepressure and temperature responses of saturated rock mass to a buried heat source at depth of about 4.8m below the ground as illustrated in Figure 7.3 (Chen, 1991; Chen *et al.*, 1991). The heat source is

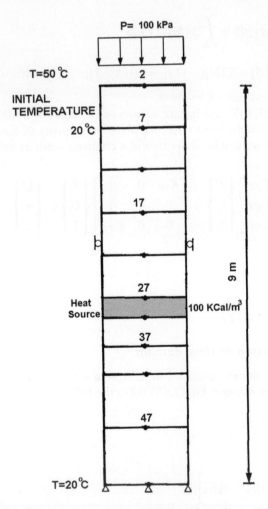

Figure 7.3 Finite element mesh and initial and boundary conditions.

$100\,\text{KCal/m}^3$ constant with time. Material properties used in the finite element analyses are given in Table 7.1.

Figures 7.4 to 7.6 show the computed responses as a function of depth at different time steps while Figures 7.7 to 7.9 show the responses of selected points as a function of time. The temperature starts to increase and it is highest at the vicinity of heat source. On the other hand, pore pressure distribution decreases as time increases. As for the displacement response, the displacement is largest at the ground surface and decreases as depth increases.

Results of the computed responses of temperature, pore pressure and displacement at selected nodes indicates that temperature tends to increase to that under steady state and pore pressure disappears as time increases. The displacements rapidly increase and tend to converge to that under steady state condition of heat field.

Table 7.1 Material properties used in computations.

Parameter	Unit	Value
Density of solid	kN/m³	25.0
Density of fluid	kN/m³	10.0
Bulk modulus of solid	N/m²	7.1×10^9
Bulk modulus of fluid	N/m²	2.3×10^9
Thermal Conductivity of solid	kW/mK	0.2
Thermal Conductivity of fluid	kW/mK	1.0
Thermal expansion coefficient of solid	1/°C	0.900×10^{-6}
Thermal expansion coefficient of fluid	1/°C	6.300×10^{-5}
Specific heat of solid	kJ/mK	14
Specific heat of fluid	kJ/mK	4
Elastic Modulus	MPa	6000
Poisson's ratio		0.280
Porosity		0.2
Hydraulic conductivity	m/day	1.86×10^{-6}

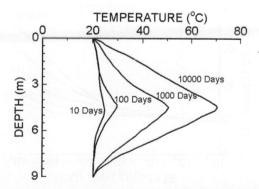

Figure 7.4 Distribution of temperature with depth at various time steps.

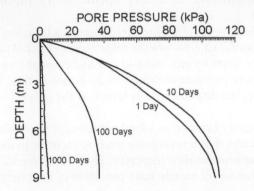

Figure 7.5 Distribution of pore pressure with depth at various time steps.

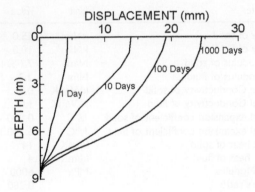

Figure 7.6 Distribution of displacement with depth at various time steps.

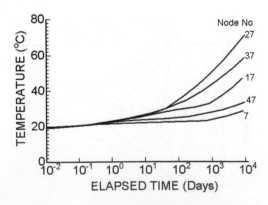

Figure 7.7 Variation of temperature at selected nodes with time.

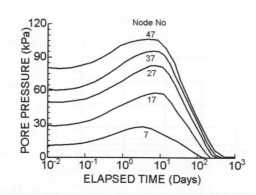

Figure 7.8 Variation of pore pressure at selected nodes with time.

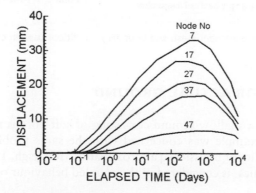

Figure 7.9 Variation of displacement at selected nodes with time.

7.4.2 Analyses of shallow and deep underground waste disposal repositories

Numerical analyses based on the simulations by Seiki (1994) and Seiki et al. (1966) are presented for shallow and deep underground, which includes a heat source, to clarify its effect on surrounding. It is also shown that Boussinesq's approximation can represent free convection.

Governing equations derived in previous section were discretized in physical space by applying Galerkin-type FEM, together with the use of 4-noded iso-parametric elements for pressure (p) and temperature (T) and 9-noded iso-parametric element for displacement (u). Then time discretization was carried out for time derivative by applying θ-method. We chose the backward difference method $(\theta = 1.0)$. To verify the numerical code, numerical analyses for coupling between heat, stress and seepage fields were carried out and it was found that numerical results agreed well with theoretical and experimental results.

In this sub-section, two examples are solved and their implications are discussed.

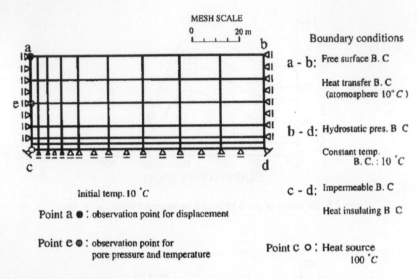

Figure 7.10 Finite element mesh and boundary conditions used in computations.

7.5 ANALYSIS FOR ACTUAL GROUND

In the first example, a shallow ground consisting of soft porous rock mass was analyzed. The second example was concerned with the same problem at great depth. Finally, the applicability of Boussinesq approximation (Rayleigh, 1962), which is often used on fluid mechanics, is considered in the coupled behaviour of rock mass.

a) Analysis for shallow ground of soft rock with a heat source

If there is a heat source underground, underground water or seepage flow induced by free convection may accelerate the diffusion of substance. Assuming that heat source with high temperature was placed in the porous ground, computations were carried out for the following two cases:

Case A: Heat transport induced by heat conduction only and
Case B: Heat transport induced by heat conduction and free convection.

The finite element mesh shown in Figure 7.10 was used. Material parameters are given in Table 7.2. To investigate the long term effect to the ground caused by the heat transport, the analysis was performed till 35,000 days or about 100 years, and time step was 25 days. Furthermore, the ground surface was set as a heat transmission boundary.

The deformations by numerical analysis are shown in Fig. 7.11 for Case A and Fig. 7.12 for Case B. These figures indicate that the displacement for Case B is larger than that for Case A. Particularly, the displacement at the point on the ground surface shows that the uplift displacement for Case B at 35,000 days was about three times

Table 7.2 Material properties used in computations.

Parameter	Unit	Value
Density of solid	t/m^3	2.167
Density of fluid	t/m^3	1.000
Bulk modulus of solid	N/m^2	9.807×10^{33}
Bulk modulus of fluid	N/m^2	2.002×10^9
Thermal Conductivity of solid	N/°C/day	1.808×10^5
Thermal Conductivity of fluid	N/°C/day	4.972×10^4
Thermal expansion coefficient of solid	1/°C	5.000×10^{-6}
Thermal expansion coefficient of fluid	1/°C	5.000×10^{-4}
Specific heat of solid	m^2/s^2/°C	1.046×10^3
Specific heat of fluid	m^2/s^2/°C	1.000×10^3
Elastic Modulus	t/m^2	1.000×10^3
Poisson's ratio		0.330
Porosity		0.444
Hydraulic conductivity	m/day	8.640×10^{-5}

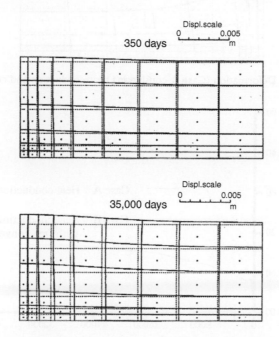

Figure 7.11 Deformation (Case A: considering heat conduction only).

larger than that for Case A (Fig. 7.13). In addition, since the ground surface was a heat transmission boundary, uplift stops when heat emitted from the heat source becomes equal to heat dissipated from the ground surface.

Fig. 7.15 shows the relations between temperature at point e and elapsed time for two cases, which are almost the same. Fig. 7.16 shows the distribution of excess pore pressure. Result for Case A shows that the excess pore pressure has a peak value

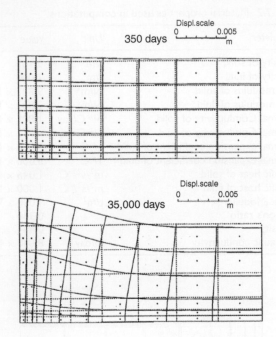

Figure 7.12 Deformation (Case B: considering heat conduction and convection).

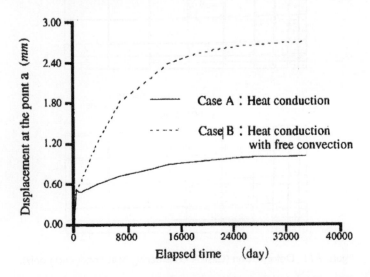

Figure 7.13 Relation between displacement at the surface and elapsed time.

immediately after applying the heat source, which was caused by the expansion of ground around the heat source and disperses. On the other hand, computed excess pore pressure for Case B increases immediately after heating and reaches an equilibrium state with time. Comparing Fig. 7.15 and Fig. 7.17, it is clear that the ground displacement is greatly influenced by excess pore pressure induced by free convection.

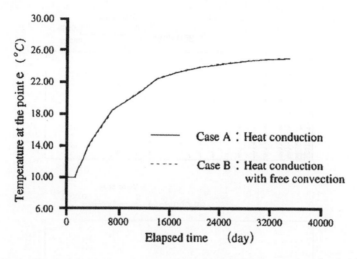

Figure 7.14 Relation between underground temperature and elapsed time.

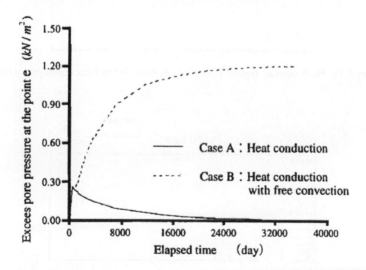

Figure 7.15 Relation between excess pore pressure of underground and elapsed time.

Distributions of flow velocity are shown in Fig. 7.16 for Case A and in Fig. 7.17 for Case B. The result for Case A (Fig. 7.16) indicates that the seepage flow in association with excess pore pressure occurs and, its velocity gets smaller as time progresses. On the other hand, the result for Case B (Fig. 7.17) indicates water flows from heat source towards the ground surface tends to return to the heat source as it cools down. As seen from the figure, the convection cell gets bigger and the velocity or seepage flow gets larger around the heat source.

In case of underground disposal of a contaminant, which generates heat, the long term migration of the seepage flow induced by free convection cannot be neglected. This is shown in the next example.

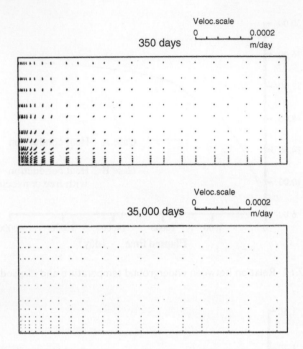

Figure 7.16 Fluid velocity distribution (Case A: considering heat conductivity only).

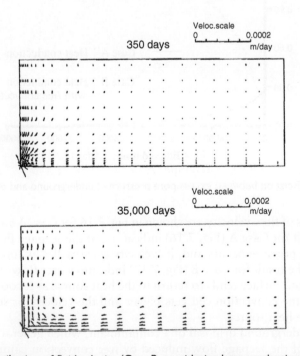

Figure 7.17 Distribution of fluid velocity (Case B: considering heat conduction and convection).

Table 7.3   Material properties used in computations different from
          Table 7.1.

Elastic Modulus	t/m²	1.000×10^6
Porosity		0.100
Hydraulic conductivity	m/day	8.640×10^{-5}

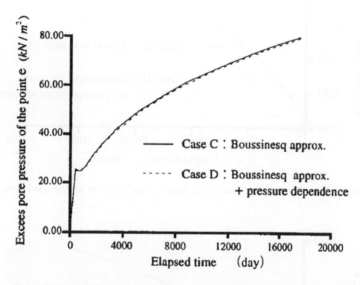

Figure 7.18   Relation between excess pore pressure and elapsed time.

b) Analysis for rockmass with a heat source at great depth

Assuming that a heat source was located in rock mass at great depth (i.e. 1,000 m),
a series of analyses was performed. The finite element mesh was the same as the
previous one shown in Fig. 7.10. Boundary conditions were set as follows: Uniformly
distributed load, which was equivalent to the overburden stress, was applied on the
upper boundary and lateral displacement was restrained at both vertical sides. The
upper side was also impermeable boundary. Temperature for initial state and at two
sides of the mesh was set as 40°C. Temperature of a heat source was set as 100°C. The
material properties different from those given in Table 7.2 are given in Table 7.3. The
time step was also the same as that of the previous case.

In this example, energy transport was assumed to take place through both
conduction and convection term. The following two cases were analyzed.

Case C: Boussinesq approximation and the change of water density induced by tem-
    perature change is considered,
Case D: Pressure dependence is added to Boussinesq approximation.

Excess pore pressure vs. elapsed time, displacement vs. elapse time and temperature vs. elapsed are shown in Fig 7.18, 7.19, and Fig. 7.20 for two cases respectively.

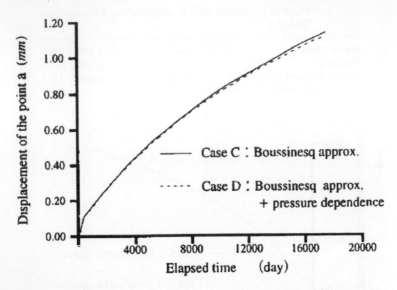

Figure 7.19 Relation between underground displacement and elapsed time.

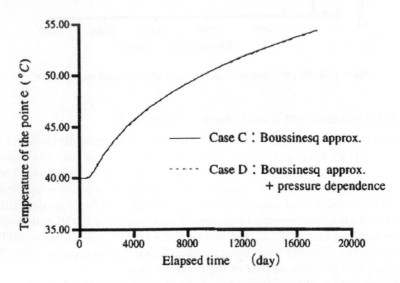

Figure 7.20 Relation between underground temperature and elapsed time.

No significant differences between the results for two cases were observed. The results also show that Boussinesq approximation can be used in the analysis of coupled behaviour of rock mass for deep disposal projects Furthermore, it is shown that although water density, generally speaking is sensitive to temperature change, it is not very sensitive to pressure or pore pressure change (Bear, 1972).

7.6   CONCLUDING REMARKS

In this chapter, a mixture theory for coupled thermo-hydro-mechanical is presented and its finite element discretization is given. Then, a series of analyses for investigating the coupled behaviour of ground at shallow and great depth were carried out and the following conclusions are drawn:

1) Considering heat transport by convection does not influence temperature distributions and no significant difference in temperature distributions were observed.
2) In case of considering heat conduction only, immediately after heating excess pore pressure gets a peak value. Simultaneously seepage flow occurs in the direction of dispersion and this flow gets smaller as time goes by.
3) In case of considering both heat conduction and convection excess pore pressure first developed due to convection. Then, its effect around the heat source gets larger with time. As a result, circulating flow occurs around the heat source and its velocity gradually gets faster.
4) As the seepage velocity caused by free convection is extremely slow, heat transmission flow does not occur. For this reason, heat transport induced by free convection is not of great significance.
5) Considering free convection, uplift displacement of ground surface is three times larger than the displacement considering heat conduction only. These results indicate that the effect of free convection could not be neglected.
6) Boussinesq approximation is also applicable to analysis of rock mass at great depth. Water density change is mainly caused by temperature change.

APPENDIX

In this Appendix, the effect of the order of shape functions is briefly investigated for a one-dimensional consolidations problem. Specifically the settlement and pore pressure variation in ground by considering is investigated for two conditions

Condition 1: The order of interpolation function of displacement field is higher than that for pore pressure
Condition 2: The order of interpolation functions for pressure and displacement field are the same.

Figures A.1 and A.2 show the variation of settlement and pore pressure at some selected nodes for the given properties shown in the figures. The displacement field was interpolated a quadratic shape function while the pressure was approximated by linear function for Condition 1. As for Condition 2, the interpolation functions for displacement and pore pressure were the same and linear. When computed responses are compared with each other, there is almost no difference among the computed response for both conditions. Theoretically the order of the displacement field should be higher than the pore pressure. Nevertheless, the computations imply that the differences are negligible. If such a conclusion is valid, this fact would be quite useful in the generation of finite element meshes.

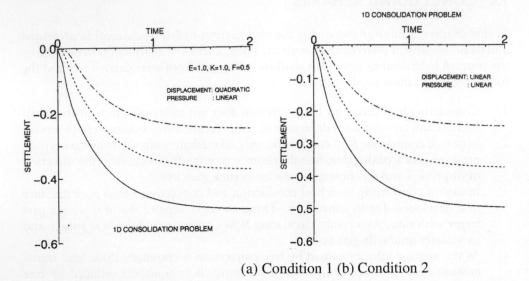

(a) Condition 1 (b) Condition 2

Figure A.1 Settlement of several selected points.

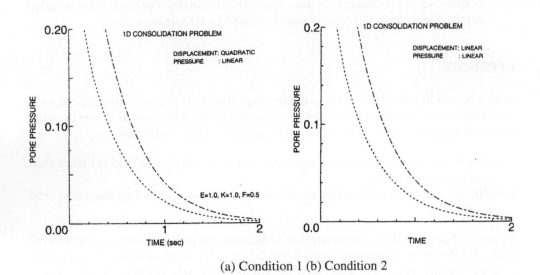

(a) Condition 1 (b) Condition 2

Figure A.2 Pore pressure of several selected points.

Chapter 8

Conclusions

The book covered a great variety of time-dependency topics in rock mechanics and rock engineering. As pointed out in the first chapter, this book includes a wide spectrum of time dependency topics in addition to the conventional concept of time-dependent behavior/rate-dependent behavior of rocks. The topics involved time-dependent behaviour of rocks, water diffusion and seepage through rocks, hydro-mechanical, thermo-mechanical, thermo-hydro-diffusion and thermo-hydro-mechanical phenomena of rocks and their applications. The fundamental procedure to explain each topic generally followed the order of theoretical formulation, experiments and practical applications.

In Chapter 2, first a brief summary of past studies is presented. Then, experimental techniques for determining the time-dependent characteristics of rocks including various constitutive models and actual experiments under various environmental conditions are described. In the final part of this chapter, several practical examples of applications to actual rock engineering structures including theoretical and numerical formulations are given with the purpose of illustrating how important it is to consider the time/rate dependency of rocks in the field of rock mechanics and rock engineering.

In Chapter 3, a theoretical method to model the water-content migration in geo-materials is described and some experimental set-ups are presented to measure the moisture migration and associated volumetric variations of geo-materials prone to water absorption/desorption. Furthermore, the experimental results are presented to show how the water content affects the physical and mechanical of properties of soft rocks, which are prone to absorption/desorption of water in their microscopic structure. The use of X-Ray scanning technique to visualize the water migration process in such rocks is explained as a non-destructive testing technique in rock mechanics. The final part of this chapter covers some important applications of the water migration phenomenon and associated issues in some specific engineering problems such as the effect of swelling/shrinkage of soft rocks on the collapse of underground openings or pressures on tunnel linings and how to evaluate the long-term creep-like response of slope stability problems due to water migration, which results in volumetric changes, causing their cracking and decomposition.

In Chapter 4, first the governing equation for heat transport in rocks and their finite element formulation are presented. Then, a brief summary for the determination of thermal properties of rocks together with an experimental technique based all-in-one concept is given. Then, temperature variations in concrete linings adjacent to rock under different thermal environment and associated thermal stress and behaviour of underground openings subjected to fluctuating thermal boundary conditions are evaluated. In the final example of applications, temperature variations in rock in the close vicinity of faults subjected to various energy release conditions are obtained and their implications discussed in practice.

In Chapter 5, the fundamental governing equations for modeling percolating fluid in rock mass and its mechanical effect are first derived using the mixture theory. Then, the theoretical formulation of Darcy's law used as a constitutive law for water perco-lation through pores of rocks and discontinuities in rock mass is presented, and the theoretical background of experimental techniques including transient pulse technique and its numerical representation are given and used to simulate some experimental results. Furthermore, some formulations and discussions are presented for evaluating the coupling effect of water on the mechanical field. In the final part of this chapter, a pumped storage hydro-electric power house together with its surrounding rock mass and upper and lower reservoirs is analyzed using the finite element method to illustrate the effect of fluctuating water levels in reservoirs on the hydro-mechanical response of surrounding rock mass.

In Chapter 6, a theoretical formulation based on the mixture theory is described for the thermo-hydro-diffusion phenomena. In the theoretical formulation, Duffour and Soret effects are considered for coupling the thermal and diffusion fields with each other and a finite element formulation of the coupled model is presented. Then numerical analyses of some laboratory tests are carried out and compared with exper-imental results. In addition, a series of parametric numerical analyses are performed to investigate

a) Simulation of solute transport in rock under laboratory conditions,
b) Temperature field of geo-thermal fields under forced seepage
c) Non-isothermal advective moisture transport through buffer materials, and
d) Parametric studies on the consideration of Duffour and Soret laws through a purely coupled hydro-thermo-diffusion formulation.

In Chapter 7, the formulation of coupled thermo-hydro-mechanical behaviour of rock masses and its finite element representation are first presented using the mixture theory. Two different examples of applications of the theory and its numerical repre-sentation of the coupled thermo-hydro-mechanical phenomenon are given. The first application involved the evaluation of displacement, pore pressure and temperature responses of saturated rock mass to a buried heat source at a very shallow depth below the ground surface. The second example of applications involved numerical analyses of heat source on the response of rock mass for shallow and deep underground condi-tions. In these numerical analyses, Boussinesq's approximation is used to consider the effect of free convection.

This book presented theoretical formulations, some experimental techniques, numerical formulations and examples of applications for a wide range of topics on time dependency in rock mechanics and rock engineering. Although this book is intended to provide a concise and advanced representation of time-dependency topics in a unified manner, it can be used a textbook for educational purposes. Chapters 2 to 5 should be appropriate for both undergraduate and graduate courses while Chapters 6 and 7 would be appropriate for graduate courses.

Appendix

Publications related to the book

TIME-DEPENDENT BEHAVIOR UNDER STRESS

Aydan, Ö. (1994) The dynamic shear response of an infinitely long visco-elastic layer under gravitational loading. *Soil Dynamics and Earthquake Engineering*, 13, 181–186.

Aydan, Ö. (1995) The stress state of the earth and the earth's crust due to the gravitational pull. *The 35th US Rock Mechanics Symposium*, Lake Tahoe, pp. 237–243.

Aydan, Ö. (1997) Dynamic uniaxial response of rock specimens with rate-dependent characteristics. In: *SARES'97.* pp. 322–331.

Aydan, Ö. (2008) New directions of rock mechanics and rock engineering: Geomechanics and geoengineering. In: *5th Asian Rock Mechanics Symposium (ARMS5), Tehran.* pp. 3–21.

Aydan, Ö. (2010) *An Experimental and Numerical Study on Creep Characteristics of Soft Rocks and Its Use for the Long-Term Response and Stability of Rock Engineering Structures.* Shibuya, Tokai University, unpublished note. 77 pp.

Aydan, Ö. (2015) The state of art on large cavern design for underground powerhouses and some long-term issues. In: *Lehr/Wiley Encyclopedia Energy: Science, Technology and Applications.* Hoboken, NJ, John Wiley and Sons.

Aydan, Ö. & Geniş, M. (2004) Surrounding rock properties and openings stability of rock tomb of Amenhotep III (Egypt). In: *ISRM Regional Rock Mechanics Symposium, Sivas.* pp. 191–202.

Aydan, Ö. & Nawrocki, P. (1998) Rate-dependent deformability and strength characteristics of rocks. In: *International Symposium on the Geotechnics of Hard Soils-Soft Rocks, Napoli,* Vol. 1. pp. 403–411.

Aydan, Ö. & Ulusay, R. (2003) Geotechnical and geoenvironmental characteristics of man-made underground structures in Cappadocia, Turkey. *Engineering Geology,* 69, 245–272.

Aydan, Ö. & Ulusay, R. (2013) Geomechanical evaluation of Derinkuyu Antique Underground City and its implications in geoengineering. *Rock Mechanics and Rock Engineering,* 46 (4), 731–754.

Aydan, Ö., Akagi, T., Okuda, H. & Kawamoto, T. (1994) The cyclic shear behaviour of interfaces of rock anchors and its effect on the long term behaviour of rock anchors. In: *International Symposium on New Developments in Rock Mechanics and Rock Engineering, Shenyang.* pp. 15–22.

Aydan, Ö., Akagi, T., Ito, T., Ito, J. & Sato, J. (1995) Prediction of deformation behaviour of a tunnel in squeezing rock with time-dependent characteristics. In: *Numerical Models in Geomechanics NUMOG V*. pp. 463–469.

Aydan, Ö., Akagi, T. & Kawamoto, T. (1996a) The squeezing potential of rock around tunnels: Theory and prediction with examples taken from Japan. *Rock Mechanics and Rock Engineering*, 29 (3), 125–143.

Aydan, Ö., Üçpırtı, H. & Kumsar, H. (1996b) The stability of a rock slope having a visco-plastic sliding surface. *Rock Mechanics Bulletin*, 12, 39–49.

Aydan, Ö., Ulusay, R., Yüzer, E. & Erdoğan, M. (1999) Man-made structures in Cappadocia, Turkey and their implications in rock mechanics and rock engineering. *ISRM News Journal*, 6 (1), 63–73.

Aydan, Ö., Sakamoto, A., Yamada, N., Sugiura, K. & Kawamoto, T. (2005a) The characteristics of soft rocks and their effects on the long term stability of abandoned room and pillar lignite mines. In: *Post Mining 2005, Nancy*.

Aydan, Ö., Daido, M., Tano, H., Tokashiki, N. & Ohkubo, K. (2005b) A real-time multi-parameter monitoring system for assessing the stability of tunnels during excavation. In: *ITA Conference, Istanbul*. pp. 1253–1259.

Aydan, Ö., Sakamoto, A., Yamada, N., Sugiura, K. & Kawamoto, T. (2005c) A real time monitoring system for the assessment of stability and performance of abandoned room and pillar lignite mines. In: *Post Mining 2005, Nancy*.

Aydan, Ö., Seiki, T., Ito, T., Ulusay, R. & Yüzer, E. (2006) A comparative study on engineering properties of tuffs from Cappadocia of Turkey and Oya of Japan. In: *Symposium on Modern Applications of Engineering Geology*. Denizli, Turkish National Group of Engineering. pp. 425–433.

Aydan, Ö., Tano, H., Geniş, M., Sakamoto, I. & Hamada, M. (2008) Environmental and rock mechanics investigations for the restoration of the tomb of Amenophis III. In: *Japan – Egypt Joint Symposium New Horizons in Geotechnical and Geoenvironmental Engineering, Tanta, Egypt*. pp. 151–162.

Aydan, Ö., Rassouli, F. & Ito, T. (2011) Multi-parameter responses of Oya tuff during experiments on its time-dependent characteristics. In: *Proceedings of the 45th US Rock Mechanics/Geomechanics Symposium*. San Francisco, CA, ARMA. pp. 11–294.

Aydan, Ö., Uehara, F. & Kawamoto, T. (2012a) Numerical study of the long-term performance of an underground powerhouse subjected to varying initial stress states, cyclic water heads, and temperature variations. *International Journal of Geomechanics, ASCE*, 12 (1), 14–26.

Aydan, Ö., Tokashiki, N. & Geniş, M. (2012b) Some considerations on yield (failure) criteria in rock mechanics ARMA 12-640. In: *46th US Rock Mechanics/Geomechanics Symposium, Chicago*, Paper No. 640. 10 pp. (on CD).

Hamada, M., Aydan, Ö. & Tano, H. (2004a) Rock mechanical investigation: Environmental and rock mechanical investigations for the conservation project in the Royal Tomb of Amenophis III. In: *Conservation of the Wall Paintings in the Royal Tomb of Amenophis III, First and Second Phases Report*. UNESCO and Institute of Egyptology, Waseda University. pp. 83–138.

Hamada, M., Aydan, Ö. & Tano, H. (2004b) Rock mechanical investigation: Responses of in-situ monitoring system at the Royal Tomb of Amenophis III during the second phase. In: *Conservation of the Wall Paintings in the Royal Tomb of Amenophis III, First and Second Phases Report*. UNESCO and Institute of Egyptology, Waseda University. pp. 221–227.

Ito, T., Aydan, Ö., Ulusay, R. & Kaşmer, Ö. (2008) Creep characteristics of tuff in the vicinity of Zelve antique settlement in Cappadocia region of Turkey. In: *5th Asian Rock Mechanics Symposium (ARMS5), Tehran*. pp. 337–344.

Sato, J., Ito, J. & Aydan, Ö. (1995) Prediction of time-dependent deformation behaviour of a tunnel in squeezing rock. In: *The 4th International Symposium on Field Measurements in Geomechanics, Bergamo*. pp. 47–54.

Tokashiki, N., Akagi, T., Ito, T. & Aydan, Ö. (2004) An experimental study on the short and long-term behavior of pillars of Ryukyu limestone. In: *3rd Asian Rock Mechanics Symposium, Kyoto*, Vol. 2. pp. 885–890.

Ulusay, R., Ito, T., Akagi, T., Seiki, T., Yüzer, E. & Aydan, Ö. (1999) Long term mechanical characteristics of Cappadocia Tuff. In: *The 9th International Rock Mechanics Congress, Paris*. pp. 687–690.

MOISTURE MIGRATION AND RELATED PHENOMENON IN ROCKS

Aydan, Ö. (2003) The moisture migration characteristics of clay-bearing geo-materials and the variations of their physical and mechanical properties with water content. In: *2nd Asian Conference on Saturated Soils, UNSAT-ASIA 2003*. pp. 383–388.

Aydan, Ö. (2008) New directions of rock mechanics and rock engineering: Geomechanics and geoengineering. In: *5th Asian Rock Mechanics Symposium (ARMS5), Tehran*. pp. 3–21.

Aydan, Ö. (2011) Some issues in tunnelling through rock mass and their possible solutions. In: *First Asian Tunnelling Conference, ATS11-15*. pp. 33–44. (Invited lecture).

Aydan, Ö. (2012) The inference of physico-mechanical properties of soft rocks and the evaluation of the effect of water content and weathering on their mechanical properties from needle penetration tests. ARMA 12-639. In: *46th US Rock Mechanics/Geomechanics Symposium, Chicago*, Paper No. 639. 10 pp. (on CD).

Aydan, Ö. & Minato, T. (2003) The swelling potential of a bentonitic buffer material and its swelling induced cracking. In: *2nd Asian Conference on Saturated Soils, UNSAT-ASIA 2003*. pp. 187–192.

Aydan, Ö. & Ulusay, R. (2003) Geotechnical and geoenvironmental characteristics of man-made underground structures in Cappadocia, Turkey. *Engineering Geology*, 69, 245–272.

Aydan, Ö., Ito, T., Akagi, T. & Kawamoto, T. (1994) Theoretical and numerical modelling of swelling phenomenon of rocks in rock excavations. In: *International Conference on Computer Methods and Advances in Geomechanics, IACMAG, Morgantown*, Vol. 3. pp. 2215–2220.

Aydan, Ö., Ulusay, R., Yüzer, E. & Erdoğan, M. (1999) Man-made structures in Cappadocia, Turkey and their implications in rock mechanics and rock engineering. *ISRM News Journal*, 6 (1), 63–73.

Aydan, Ö., Sakamoto, A., Yamada, N., Sugiura, K. & Kawamoto, T. (2005) The characteristics of soft rocks and their effects on the long term stability of abandoned room and pillar lignite mines. In: *Post Mining 2005, Nancy.*

Aydan, Ö., Seiki, T., Ito, T., Ulusay, R. & Yüzer, E. (2006) A comparative study on engineering properties of tuffs from Cappadocia of Turkey and Oya of Japan. In: *Symposium on Modern Applications of Engineering Geology, Turkish National Group of Engineering, Denizli.* pp. 425–433.

Aydan, Ö., Tano, H., Ulusay, R. & Jeong, G.C. (2008) Deterioration of historical structures in Cappadocia (Turkey) and in Thebes (Egypt) in soft rocks and possible remedial measures. In: *2008 International Symposium on Conservation Science for Cultural Heritage, Seoul.* pp. 37–41.

Aydan, Ö., Rassouli, F. & Ito, T. (2011) Multi-parameter responses of Oya tuff during experiments on its time-dependent characteristics. In: *Proceedings of the 45th US Rock Mechanics/Geomechanics Symposium.* San Francisco, ARMA. pp. 11–294.

Aydan, Ö., Sato, A. & Yagi, M. (2014) The inference of geo-mechanical properties of soft rocks and their degradation from needle penetration tests. *Rock Mechanics and Rock Engineering*, 47, 1867–1890.

Didier, C., Van Der Merwe, N., Betournay, M., Mainz, M., Aydan, O., Kotyrba, A., Song, W.K. & Josien, J.P. (2009) Presentation of the ISRM mine closure state of the art report. In: *Proceedings of the ISRM-Sponsored International Symposium on Rock Mechanics: "Rock Characterisation, Modelling and Engineering Design Methods" (SINOROCK 2009), 19–22 May 2009, Hong Kong, China.*

Hamada, M., Aydan, Ö. & Tano, H. (2004a) Rock mechanical investigation: Environmental and rock mechanical investigations for the conservation project in the Royal Tomb of Amenophis III. In: *Conservation of the Wall Paintings in the Royal Tomb of Amenophis III, First and Second Phases Report.* UNESCO and Institute of Egyptology, Waseda University. pp. 83–138.

Hamada, M., Aydan, Ö. & Tano, H. (2004b) Rock mechanical investigation: Responses of in-situ monitoring system at the Royal Tomb of Amenophis III during the second phase. In: *Conservation of the Wall Paintings in the Royal Tomb of Amenophis III, First and Second Phases Report.* UNESCO and Institute of Egyptology, Waseda University. pp. 221–227.

Kano, K., Doi, T., Daido, M. & Aydan, Ö. (2004) The development of electrical resistivity technique for real-time monitoring and measuring water-migration and its characteristics of soft rocks. In: *3rd Asian Rock Mechanics Symposium, Kyoto*, Vol. 2. pp. 851–854.

Kumsar, H., Aydan, Ö., Çobanoğlu, İ., Çelik, S.B. & Akgün, M. (2012a) Investigation of the factors causing slope instabilities on the left abutment of Denizli-Gökpınar Dam. In: *The 65th Turkish Geogical Congress, Ankara.*

Kumsar, H., Çelik, S., Aydan, Ö., Tano, H. & Ulusay, R. (2012b) Investigation of Babadağ Gündoğdu landslide by using multi-parameter monitoring techniques and its evaluation within natural hazard scope. In: *Prof. Dr. Mahir Vardar Special Sessions on Geomechanics, Tunnelling, Design of Rock Constructions, Istanbul.* pp. 311–337.

Otsuka, N., Tsuruhara, H., Man & Aydan, Ö. (1990) Deformation behaviour of swelling ground and its prediction (in Japanese). In: *The 21st Rock Mechanics Symposium of Japan*. pp. 36–41.

Seiki, T. & Aydan, Ö. (2003a) Mechanical property changes of Oya tuff due to deterioration and a consideration for its effect on on the stability of an underground quarry in Oya tuff. In: *International Colloquium on Instrumentation and Monitoring of Landslides and Earthquakes in Japan and Turkey*. pp. 27–38.

Seiki, T. & Aydan, Ö. (2003b) Deterioration of Oya tuff and its mechanical property change as building stone. In: *International Symposium on Industrial Minerals and Building Stones, Istanbul*. pp. 329–336.

Ulusay, R. & Aydan, Ö. (2011) Issues on short- and long-term stability of historical and modern man-made cavities in the Cappadocia Region of Turkey. In: *First Asian Tunnelling Conference, ATS11-21*. pp. 72–83. (Invited lecture).

Ulusay, R., Aydan, Ö., Geniş, M. & Tano, H. (2013) Stability assessment of Avanos underground congress centre (Cappadocia, Turkey) in soft tuffs through an integrated scheme of rock engineering methods. *Rock Mechanics and Rock Engineering*, 46, 1303–1321.

THERMO-MECHANICAL BEHAVIOR OF ROCKS AND HEAT TRANSPORT IN ROCKS

Aydan, Ö. (1994) Thermo-mechanical performance of thick concrete linings cast against frozen rock mass during the hydration of cement. In: *International Conference on Computational Methods in Structural and Geotechnical Engineering, Hong Kong*. pp. 441–446.

Aydan, Ö. & Ersen, A. (1983) Ground-water freezing method and its application (in Turkish). *Madencilik*, 22 (2), 33–44.

Aydan, Ö., Uehara, F. & Kawamoto, T. (2012) Numerical study of the long-term performance of an underground powerhouse subjected to varying initial stress states, cyclic water heads, and temperature variations. *International Journal of Geomechanics, ASCE*, 12 (1), 14–26.

Aydan Ö., Ersen, A., Ichikawa, Y. & Kawamoto, T. (1985) Temperature and thermal stress distributions in mass concrete shaft and tunnel linings during the hydration of concrete (in Turkish). In: *The 9th Mining Science and Technology Congress of Turkey, Ankara*. pp. 355–368.

Aydan, Ö., Güloğlu, R. & Kawamoto, T. (1986) Temperature distributions and thermal stresses in tunnel linings due to hydration of cement (in Japanese). *Tunnels and Underground*, 17 (2), 29–36.

Aydan, Ö., Manav, H., Yaoita, T. & Yagi, M. (2014) Multi-parameter thermodynamic response of minerals and rocks during deformation and fracturing. In: *Proceedings of the 8th Asian Rock Mechanics Symposium, Sapporo*, 817–826.

Aydan, Ö., Fuse, T. & Ito, T. (2015) An experimental study on thermal response of rock discontinuities during cyclic shearing by Infrared (IR) thermography. In: *Proceedings of the 43rd Symposium on Rock Mechanics, JSCE*, 123–128.

HYDRO-MECHANICAL BEHAVIOR OF ROCKS AND RELATED ISSUES

Aydan, Ö. (1998a) A data-base system for seepage characteristics of geomaterials. In: *International Symposium on Environemental Issues and Waste Management in Energy and Mineral Production, Ankara.* pp. 209–213.

Aydan, Ö. (1998b) Finite element analysis of transient pulse method tests for permeability measurements. In: *The 4th European Conference on Numerical Methods in Geotechnical Engineering-NUMGE98, Udine.* pp. 719–727.

Aydan, Ö. (1998c) The effect of compressibility of sample on permeability measurements by transient pulse method. In: *The 4th National Rock Mechanics Symposium of Turkey, Zonguldak.* pp. 95–102.

Aydan, Ö. (1999) Simple tests on the validation of the effective stress law for rock discontinuities. In: *Poromechanics, Louvain-la-Neuve.* pp. 539–544.

Aydan, Ö. (2003) The mechanism of the long-term landslide at Babadağ. In: *International Colloquium on Instrumentation and Monitoring of Landslides and earthquakes in Japan and Turkey.* pp. 39–50.

Aydan, Ö. (2012) The inference of physico-mechanical properties of soft rocks and the evaluation of the effect of water content and weathering on their mechanical properties from needle penetration tests. ARMA 12-639. In: *46th US Rock Mechanics/Geomechanics Symposium, Chicago,* Paper No. 639. 10 pp. (on CD).

Aydan, Ö. & Üçpırtı, H. (1997) The theory of permeability measurement by transient pulse test and experiments. *Journal of the School of Marine Science and Technology,* 43, 45–66.

Aydan, Ö., Üçpırtı, H. & Ulusay, R. (1997a) Theoretical formulation of Darcy's law for fluid flow through porous and/or jointed rock and its validity (in Turkish). *Rock Mechanics Bulletin,* 13, 1–18.

Aydan, Ö., Üçpırtı, H. & Türk, N. (1997b) Theory of laboratory methods for measuring permeability of rocks and tests. *Rock Mechanics Bulletin,* 13, 19–36.

Aydan, Ö., Uehara, F. & Kawamoto, T. (2012) Numerical study of the long-term performance of an underground powerhouse subjected to varying initial stress states, cyclic water heads, and temperature variations. *International Journal of Geomechanics, ASCE,* 12 (1), 14–26.

Didier, C., Van Der Merwe, N., Betournay, M., Mainz, M., Aydan, O., Kotyrba, A., Song, W.K. & Josien, J.P. (2009) Presentation of the ISRM mine closure state of the art report. In: *Proceedings of the ISRM-Sponsored International Symposium on Rock Mechanics: "Rock Characterisation, Modelling and Engineering Design Methods" (SINOROCK 2009), 19-22 May 2009, Hong Kong, China.*

Shimizu, Y., Tatematsu, H. & Aydan, Ö. (1998) The relation between the roughness and permeability of rock discontinuities. In: *The 10th National Rock Mechanics of Japan, Kyoto.* pp. 97–102.

Tokashiki, N. & Aydan, Ö. (2010) Kita-Uebaru natural rock slope failure and its back analysis. *Environmental Earth Sciences,* 62 (1), 25–31.

Üçpırtı, H. & Aydan, Ö. (1997) A data-base system for seepage characteristics of intact rocks, discontinuities and rock masses. In: *Rock Mechanics and Environmental Geotechnology – RMEG.* pp. 51–56.

THERMO-HYDRO-DIFFUSION BEHAVIOR OF ROCKS

Aydan, Ö. (2001a) A finite element method for fully coupled hydro-thermo-diffusion problems and its applications to geo-science and geo-engineering. In: *10th IACMAG Conference, Austin.* pp. 781–786.

Aydan, Ö. (2001b) Modelling and analysis of fully coupled hydro-thermo-diffusion phenomena. In: *International Symposium on Clay Science for Engineering.* Is-Shizuoka, Balkema. pp. 353–360.

Aydan, Ö. (2008) New directions of rock mechanics and rock engineering: Geomechanics and geoengineering. In: *5th Asian Rock Mechanics Symposium (ARMS5), Tehran.* pp. 3–21.

Cheng, C., Aydan, Ö., Ichikawa, Y. & Kawamoto, T. (1991) Thermo-hydro-mechanical coupled analysis of ground using the mixture theory (in Japanese). In: *The 46th Annual Meeting of Japan Society of Civil Engineers,* Vol. III-473. pp. 968–969.

Seiki, T., Ichikawa, Y. & Aydan, Ö. (1996) Coupled convective heat, stress and seepage behaviour of porous rock. In: *NARMS'96.* pp. 1321–1328.

References

Bakhvalov, N. & Panasenko, G. (1984) *Homogenization: Averaging Processes in Periodic Media*. Dordrecht, Kluwer Academic Publications.

Barron, K., Hedley, D.G.F. & Coates, D.F. (1970) Field instrumentation for rock slopes. In: *Proceedings of the 1st Int. Conference Stability in Open Pit Mining. Vancouver, Canada*. pp. 143–168.

Basha, H.A. & Selvadurai, A.P.S. (1998) Heat induced moisture transport in the vicinity of a spherical heat source. *International Journal for Numerical and Analytical Methods in Geomechanics*, 22, 969–981.

Bear, J. (1988) *Dynamics of Fluids in Porous Media*. New York, NY, Elsevier.

Bieniawski, Z.T. (1970) Time-dependent behaviour of fractured rock. *Rock Mechanics*, 2, 123–137.

Biot, M.A. (1941) General theory of three-dimensional consolidation. *Journal of Applied Physics*, 12, 155–164.

Biot, M.A. (1942) The general theory of three dimensional consolidation. *Journal of Applied Physics*, 12, 155–165.

Biot, M.A. (1956a) Mechanics of deformation and acoustic propagation in porous media. *Journal of Applied Physics*, 27, 240–253.

Biot, M.A. (1956b) Theory of propagation of elastic waves in a fluid saturated porous solid: Part-I, Low-frequency range. *The Journal of the Acoustical Society of America*, 28, 168–178.

Biot, M.A. (1956c) Thermoelasticity and irreversible thermodynamics. *The Journal of Applied Physics*, 27 (3), 240–253.

Birch, F. (1950) Flow of heat in the Front Range. Colorado. *Bulletin of the Geological Society of America*, 61, 567–630.

Boukharov, G.N., Chandi, M.W. & Boukharov, N.G. (1995) The three processes of brittle crystalline rock creep. *International Journal of Rock Mechanics and Mining Science & Geomechanics Abstracts*, 32 (4), 325–335.

Brace, W.F., Walsh, J.B. & Frangos, W.T. (1968) Permeability of granite under high pressure. *Journal of Geophysical Research*, 73 (6), 2225–2236.

Brekke, T.L. (1965) On the measurement of the relative potential swellability of hydrothermal montmorillonite clay from joints and faults in pre-cambrian and paleozoic rocks in Norway. *International Journal of Rock Mechanics and Mining Sciences*, 2, 155–165.

Broch, E. (1979) Changes in rock strength by water. In: *Proceedings of the IV. International Society of Rock Mechanics, Montreux*, Vol. 1. pp. 71–75.

Byerlee, J.D. (1967) Frictional characteristics of granite under high confining pressure. *Journal of Geophysical Research*, 38, 3639–3648.

Carslaw, H. & Jaeger, J. (1959) *Conduction of Heat in Solids*. 2nd edition. Oxford, Oxford University Press. 510 pp.

Chen, J.H. (1991) *Mechanical Properties of Discontinuous Rock Mass and Thermo-Hydro-Mechanical Coupling Problem*. Master Thesis. Nagoya, Nagoya University. 30 pp.

Chen, C., Aydan, Ö., Ichikawa, Y. & Kawamoto, T. (1991) Thermo-hydro-mechanical coupled analysis of ground using the mixture theory (in Japanese). In: *The 46th Annual Meeting of Japan Society of Civil Engineers*, Vol. III-473. pp. 968–969.

Clark Jr., S.P. (1966) *Handbook of Physical Constants*. New York, NY, Geological Society of America Memoir 97. 587 pp.

Colback, P.S.B. & Wiid, B.L. (1965) The influence of moisture content on the compressive strength of rock. In: *Proceedings of the 3rd Canadian Rock Mechanics Symposium*. pp. 65–83.

Cristescu, N.D. & Hunsche, U. (1998) *Time Effects in Rock Mechanics*. New York, NY, John Wiley and Sons. 343 pp.

Doktan, M. (1983) *The Longterm Stability of Room and Pillar Workings in a Gypsum Mine*. PhD Thesis. Newcastle Upon Tyne, Newcastle University

Duncan, N., Dunne, M.H. & Petty, S. (1968) Swelling characteristics of rock. *Water Power*, 185–192.

Eringen, A.C. (1980) *Mechanics of Continua*. 2nd edition. Huntington, NY, R.E. Krieger Publishing Company.

Eringen, A.C. & Ingram, J.D. (1965) A continuum theory of chemically reacting media-I. *International Journal of Engineering Science*, 3, 197–212.

Franklin, J.A. & Chandra, R. (1972) The slake durability test. *International Journal of Rock Mechanics and Mining Science & Geomechanics Abstracts*, 9, 325–341.

Franklin, J.A. & Dusseault, M.B. (1989) *Rock Engineering*. New York, NY, McGraw Hill Publication Company.

Gale, J. (1990) Hydraulic behaviour of rock joints. In: *International Symposium on Rock Joints, Loen*. pp. 351–362.

Garg, S.K. (1971) Wave propagation effects in a fluid-saturated porous solid. *Journal of Geophysical Research*, 76, 7947–7962.

Garg, S.K. & Nur, A. (1973) Effective stress laws for fluid saturated porous rocks. *Journal of Geophysical Research*, 78, 5911–5920.

Green, A.E. & Naghdi, P.M. (1969) On basic equations for mixtures. *The Quarterly Journal of Mechanics and Applied Mathematics*, 22, 427–438.

Gustaffson, S.E. (1991) Transient plane source techniques for thermal conductivity and thermal diffusivity measurements of solid materials. *Review of Scientific Instruments*, 62, 797–804.

Hoek, E. & Brown, E.T. (1980) *Underground Excavations in Rock*. London, The Institution of Mining and Metallurgy.

Hondros, G. (1959) The evaluation of Poisson's ratio and the modulus of materials of low tensile resistance by the Brazilian (indirect tensile) tests with particular reference to concrete. *Australian Journal of Applied Sciences*, 10, 243–268.

Hubbert, M.K. & Rubey, W.W. (1959) Role of fluid pressure in mechanics of overthrust faulting. *Geological Society of America Bulletin*, 70, 115–166.

Hunsche, U. (1992) True triaxial failure tests on cubic rock salt samples – Experimental methods and results. In: *Proceedings of the IUTAM Symposium on Finite Inelastic Deformations – Theory and Applications*. Hannover, Springer Verlag. pp. 525–538.

Huyakorn, P.S. & Pinder, G.F. (1983) *Computational Methods in Subsurface Flow*. London, Academic Press. 473 pp.

Hyde, T.H., Sun, W. & Becker, A.A. (1996) Analysis of the impression creep test method using a rectangular indenter for determining the creep properties in welds. *International Journal of Mechanical Science*, 38, 1089–1102.

Igarashi, S. & Tanaka, Y. (1998) The effect of a discontinuity on solute transport through sandstone. In: *10th National Rock Mechanics Symposium, Tokyo, JRMS*. pp. 359–364.

Ishizuka, Y., Koyama, H. & Komura, S. (1993) Effect of strain rate on strength and frequency dependence of fatigue failure of rocks. In: *Proceedings of Assessment and Prevention of Failure Phenomena in Rock Engineering*. pp. 321–327.

ISRM (International Society for Rock Mechanics) (1981) *Rock Characterization Testing and Monitoring: ISRM Suggested Methods*. Brown, E.T. (ed.) Oxford, Pergamon Press. 211 pp.

Ito, H. & Sasajima, S. (1980) Long term creep experiment on some rocks observed over three years. *Tectonophysics*, 62 (3–4), 219–232.

Ito, H. & Sasajima, S. (1987) A ten year creep experiment on small rock specimens. *International Journal of Rock Mechanics and Mining Science & Geomechanics Abstracts*, 24, 113–121.

Ito, T. & Akagi, T. (2001) Methods to predict the time of creep failure. In: *Proceedings of the 31st Symposium on Rock Mechanics of Japan*. pp. 77–81.

Ito, T., Aydan, Ö., Ulusay, R. & Kaşmer, Ö. (2008) Creep characteristics of tuff in the vicinity of Zelve antique settlement in Cappadocia region of Turkey. In: *5th Asian Rock Mechanics Symposium (ARMS5), Tehran*. pp. 337–344.

Jaeger, J.C. & Cook, N.G.W. (1979) *Fundamentals of Rock Mechanics*. 3rd edition. London, Chapman & Hall. p. 79, 311.

Jones, R.M. (1975) *Mechanics of Composite Materials*. Washington, DC, McGraw-Hill Book Company.

Kanno, T., Fujita, T., Takeuchi, S., Ishikawa, H., Hara, K. & Nakano, M. (1999) Coupled thermo-hydro-mechanical modelling of bentonite buffer material. *International Journal for Numerical and Analytical Methods in Geomechanics*, 23, 1281–1307.

Kano, K., Doi, T., Daido, M. & Aydan, Ö. (2004) The development of electrical resistivity technique for real-time monitoring and measuring water-migration and its characteristics of soft rocks. In: *Proceedings of 4th Asia Rock Mechanics Symposium, Kyoto*. pp. 851–854.

Karaca, M., Aydan, Ö. & Sezaki, M. (1995) Assessing the mechanical response of rock structures subject to water forces. In: *Proceedings of the 35th US Rock Mechanics Symposium, Lake Tahoe.* pp. 335–340.

Kawamoto, T., Aydan, Ö. & Tsuchiyama, S. (1991) A consideration on the local instability of large underground openings. In: *International Conference on Geomechanics'91, Hradec.* pp. 33–41.

Ketcham, R.A. & Carlson, W.D. (2001) Acquisition, optimization and interpretation of X-ray computed tomographic imagery. Applications to the geosciences. *Computers and Geosciences,* 27, 381–400.

Kovari, K. & Fritz, P. (1975) Stability analysis of rock slopes for plane and wedge failure with the aid of a programmable pocket calculator. In: *Proceedings of the 16th US Rock Mechanics Symposium.* pp. 25–33.

Kreyszig, E. (1983) *Advanced Engineering Mathematics.* New York, NY, John Wiley & Sons.

Kumsar, H., Aydan, Ö. & Sakoda, S. (1997) Model wedges tests and re-assessment of limiting equilibrium methods for wedge sliding. *Rock Mechanics and Environmental Geotechnology – RMEG,* 261–266.

Kumsar, H., Aydan, Ö., Tano, H., Çelik, S.B. & Ulusay, R. (2016) An integrated geomechanical investigation, multi-parameter monitoring and analyses of Babadağ-Gündğdu creep-like landslide. *Rock Mechanics and Rock Engineering,* 49 (6), 2277–2299.

Ladanyi, B. (1974) Use of the long term strength concept in the determination of ground pressure on tunnel linings. In: *3rd Congr. Int. Soc. Rock Mech., Denver,* Vol. 2B pp. 1150–1165.

Ladanyi, B. (1993) Time-dependent response of rock around tunnels. In: *Comprehensive Rock Engineering,* Vol. 2 Amsterdam, Elsevier. pp. 77–112.

Lomnitz, C. (1956) Creep measurements in igneous rocks. *Journal of Geology,* 64, 473–479.

Maeda, N., Sato, T., Matsui, H. & Sugihara, K. (1999) Estimation of applicability of stress measurement methods and three dimensional stress state in soft sedimentary rocks. In: *Proceedings of 99 Japan-Korea Joint Rock Engineering Symposium, Aug. 2–4, Fukuoka, Japan.* pp. 277–284.

Mamaghani, I.H.P., Baba, S., Aydan, Ö. & Shimizu, S. (1994) Discrete finite element method for blocky systems. In: *Proceedings of Computer Methods and Advances in Geomechanics, Morgantown,* Vol. 1. pp. 843–850.

Marsden, J.R., Holt, R.M., Nakken, S.J. & Raaen, A.M. (1992) Mechanical and petrophysical characterization of highly stressed mudstones. In: *EUROCK 92.* London, Thomas Telford. pp. 51–56.

Masuda, K., Mizutani, H. & Yamada, I. (1987) Experimental study of strain-rate dependence and pressure dependence of failure properties of granite. *Journal of Physics of the Earth,* 35, 37–66.

Masuda, K., Mizutani, H., Yamada, I. & Imanishi, Y. (1988) Effects of water on time-dependent behavior of granite. *Journal of Physics of the Earth,* 36, 291–313.

Morland, L.W. (1972) A simple constitutive theory for a fluid-saturated porous solid. *Journal of Geophysical Research,* 77 (5), 891–900.

Mottahed, P. & Szeki, A. (1982) *The Collapse of Room and Pillar Workings in a Shaley Gypsum Mine Due to Dynamic Loading Symp. on Strata Mech. Newcastle.* pp. 260–264.

Murayama, S. & Yagi, N. (1966) Swelling of mudstone due to sucking of water. In: *Proceedings of 1st Int. Congr. Rock Mechanics, ISRM, Lisbon*, Vol. 1. pp. 452–455.

Nikolaevsky, V.N. (1985) *Mechanics of Fluid-Saturated Geomaterials: Mechanics of Geomaterials.* Bazant, Z. (ed.). New York, NY, John Wiley and Sons Ltd. pp. 379–401.

Okubo, S., Nishimatsu, Y. & Fukui, K. (1991) Complete creep curves under uniaxial compression. *International Journal of Rock Mechanics and Mining Science & Geomechanics Abstracts*, 28, 77–82.

Okubo, S., Fukui, K. & Nishimatsu, Y. (1993) Control performance of servocontrolled testing machines in compression and creep tests. *International Journal of Rock Mechanics and Mining Science & Geomechanics Abstracts*, 30, 247–255.

Owen, D.R.J. & Hinton, E. (1980) *Finite Element in Plasticity: Theory and Practice.* Swansea, Pineridge Press Ltd.

Özkol, S. (1965) *Swelling Characteristics of Permian Clay.* PhD Thesis. Stillwater, OK, Oklahoma State University.

Parker, W., Jenkins, R., Butler, C. & Abbott, G. (1961) Flash method of determining thermal diffusivity, heat capacity, and thermal conductivity. *Journal of Applied Physics*, 32, 1679–1685.

Pasamehmetoglu, A.G., Bozdag, T. & Yesil, M.M. (1993) The three dimensional swelling behaviour of clay bearing rocks. In: *Proceedings of the International Symposium on Assessment and Prevention of Failure Phenomena in Rock Engineering.* pp. 357–360.

Peng, S. (1973) Time-dependent aspects of rock behavior as measured by a servo-controlled hydraulic testing machine. *International Journal of Rock Mechanics and Mining Science & Geomechanics Abstracts*, 10, 235–246.

Perzyna, P. (1966) Fundamental problems in viscoplasticity. *Advances in Applied Mechanics*, 9 (2), 244–368.

Pfister, M., Rybach, L. & Şimşek, Ş. (1997) Geothermal reconnaissance of the Marmara Sea Region (NW Turkey). In: *Active Tectonics of Northwestern Anatolia-The Marmara Poly-Project.* Zurich, vdf Hochschulverlag AG an der ETH Zürich. pp. 503–535.

Polubarinova-Kochina, P.YA. 1962. *Theory of Groundwater Movement.* Princeton, Princeton University Press.

Popov, Y., Pribnow, D., Sass, J., Williams, C. & Burkhardt, H. (1999) Characterisation of rock thermal conductivity by high-resolution optical scanning. *Geothermics*, 28, 253–276.

Rayleigh, L. (1962) On convection currents in a horizontal layer of fluid, when the higher temperature is on the under side. In: Saltzman, B. (ed.) *Selected Papers on the Theory of Thermal Convection with Special Application to Earth's Planetary Atmosphere.* Mineola, NY, Dover. pp. 3–20.

Sass, J., Stone, C. & Munroe, R. (1984) Thermal conductivity determinations on solid rock – A comparison between a steady-state divided-bar apparatus and a commercial transient line-source device. *Journal of Volcanology and Geothermal Research*, 20 (1–2), 145–153.

Sato, A. & Aydan, Ö. (2014) An X-Ray CT imaging of water absorption process of soft rocks. In: *Unsaturated Soils: Research and Applications – Proceedings of the 6th International Conference on Unsaturated Soils, UNSAT 2014.*

Sato, A., Kataoka, M. & Obara, Y. (2011) Analysis of tracer diffusion process trough crack surfaces in porous rock by means of X ray CT. In: *Proceedings of the ROCMEC'2011-Xth Regional Rock Mechanics Symposium, Ankara, Turkey.* pp. 187–194.

Scholz, C.H. (1968) Mechanism of creep in brittle rock. *Journal of Geophysical Research*, 73, 3295.

Seiki, T. (1994) *Mechanical Properties of Discontinuous Rock Mass and Thermo-Hydro-Mechanical Coupling Problem.* PhD Thesis. Nagoya, Nagoya University. 125 pp.

Seiki, T., Ichikawa, Y. & Aydan, Ö. (1996) Coupled convective heat, stress and seepage behaviour of porous rock. In: *NARMS'96.* pp. 1321–1328.

Snow, D.T. (1965) *A Parallel Plate Model of Fractured Permeable Media.* PhD Dissertation. Berkeley, University of California.

Somerton, W.H. (1992) *Thermal Properties and Temperature-Related Behavior of Rock/Fluid Systems. Developments in Petroleum Science*, Vol. 37. Amsterdam, Elsevier Science Publishers B.V. 257 pp.

Temel, A. (2002) *Personal Communication.* Ankara, Geological Engineering Department, Hacettepe University.

Terzaghi, K. (1925) *Erdbaumechanik auf bodenphysikalischer Grundlage.* Leipzig, Vienna, F. Deuticke's Verlag.

Trusdell, C.A. (1969) *Rational Thermodynamics.* New York, NY, McGraw-Hill Series in Modern Applied Mathematics. pp. 81–98.

Trusdell, C.A. & Toupin, R.A. (1960) The classical field theories. In: Flügge, S. (ed.) *Handbuch der Physik III/1.* Berlin, Springer Verlag. pp. 226–793.

Tsang, Y.W. & Witherson, P.A. (1981) Hydro-mechanical behaviour of a deformable rock fracture subject to normal stress. *Journal of Geophysical Research*, 86 (B10), 9287–9298.

Üçpırtı, H. & Aydan, Ö. (1997) An experimental study on the permeability of interface between sealing plug and rock. In: *The 28th Rock Mechanics Symposium of Japan.* pp. 268–272.

Üçpırtı, H., Daemen, J.J.K., Finley, R.E. & George, J.T. (1992) *Laboratory Measurement of Gas Flow Along a Pressurized Grout/Membrane/Halite Interface for Waste Isolation Pilot Plan.* SAND92-2121. Albuquerque, NM, Sandia National Laboratories.

Vardar, M. & Fecker, E. (1986) Examples from problematic projects involving evaporates and fundamentals of dealing such problems. In: *1st National Rock Mechanics Symposium of Turkey, Ankara.* pp. 289–310 (in Turkish).

Watanabe, H., Tano, H., Ulusay, R., Yüzer, E., Erdoğan, E. & Aydan, Ö. (1999) The initial stress state in Cappadocia. In: Matsui, K. & Shimada, H. (eds.) *Proceedings of the '99 Japan-Korea Joint Symposium on Rock Engineering, Fukuoka, Japan.* pp. 249–260.

Wawersik, W.R. (1972) Time-dependent rock behavior in uniaxial compression. In: *Proceedings of the 14th Symposium on Rock Mechanics*. University Park, PA, Pennsylvania State University. pp. 85–106.

Wawersik, W.R. (1983) Determination of steady state creep rates an activation parameters for rock salt. In: *High Pressure Testing of Rock*. Special Technical Publication of ASTM, STP869. pp. 72–91.

Wittke, W. (1990) *Rock Mechanics*. Berlin, Springer-Verlag.

Yesil, M.M., Pasamehmetoglu, A.G. & Bozdag, T. (1993) A triaxial swelling test apparatus. *International Journal of Rock Mechanics and Mining Sciences*, 30, 443–450.

Zanbak, C. & Arthur, R.C. (1984) Rock mechanics aspects of volume changes in calcium sulfate bearing rocks due to geochemical phase transitions. In: *Proceedings of the 25th US Symposium on Rock Mechanics*. pp. 328–337.

Zimmerman, R.W., Somerton, W.H. & King, M.S. (1986) Compressibility of porous rocks. *Journal of Geophysical Research*, 91 (B12), 12765–12777.

Subject index

ISRM Book Series

Book Series Editor: Xia-Ting Feng

ISSN: 2326-6872

Publisher: CRC Press/Balkema, Taylor & Francis Group

1. Rock Engineering Risk
   Authors: John A. Hudson & Xia-Ting Feng
   2015
   ISBN: 978-1-138-02701-5 (Hbk)

2. Time-Dependency in Rock Mechanics and Rock Engineering
   Author: Ömer Aydan
   2016
   ISBN: 978-1-138-02863-0 (Hbk)

Printed and bound by CPI Group (UK) Ltd, Croydon, CR0 4YY

23/10/2024

01778379-0002